Foundations of NANOMECHANICS

Foundations of NANOMECHANICS

by

Antara Raut | Rasika Gupta | Diya Pandit

Wordpen Academics

Publishing Science That Matter

ISBN: 978-93-6084-191-1 (Hardback)

Published by : **WORDPEN ACADEMICS**
4736/23, Ansari Road, Darya Ganj,
New Delhi – 110 002
Phone: +91-011-35004336
E-mail: info@wordpenacademics.com

Printed at **:** **Replika Press Pvt. Ltd.**

Exclusive Distributor: ***Bio-Green Books, New Delhi***

Preface

In the realm of nanotechnology, nanomechanics stands as a frontier of exploration, offering insights into the mechanical behavior of materials and structures at the nanoscale. From the bending of carbon nanotubes to the stretching of DNA strands, nanomechanics unveils the fascinating world of small-scale phenomena, where the laws of classical mechanics intersect with quantum effects and surface forces. As scientists and engineers delve into the foundations of nanomechanics, they uncover new possibilities for designing innovative materials, devices, and systems with unprecedented mechanical properties and functionalities.

This book, titled "Foundations of Nanomechanics," serves as a comprehensive guide to understanding the principles, theories, and applications of nanomechanics in the field of nanotechnology. Drawing upon the expertise of physicists, materials scientists, and engineers, this volume offers insights, methodologies, and practical approaches for studying and manipulating mechanical behavior at the nanoscale.

From the basics of continuum mechanics and solid-state physics to the intricacies of atomistic simulations and scanning probe techniques, "Foundations of Nanomechanics" covers a wide range of topics essential for mastering the principles of nanomechanics. Through theoretical explanations, mathematical derivations, and computational simulations, readers will gain a deep understanding of the fundamental principles governing mechanical behavior at the nanoscale, including elasticity, plasticity, fracture, and surface interactions.

Moreover, this book explores the applications of nanomechanics in various fields, from nanoelectronics and nanomedicine to nanomaterials and nanomanufacturing. By showcasing real-world examples, case studies, and emerging trends, we highlight the potential impact of nanomechanics on technology development and innovation, paving the way for breakthroughs in areas such as nanoscale sensors, actuators, and biomaterials.

As we embark on this journey of exploration and discovery, it is our hope that "Foundations of Nanomechanics" will serve as a valuable resource for students, researchers, and practitioners in the field of nanotechnology. Whether you are a graduate student delving into the intricacies of nanomechanics or a seasoned researcher seeking to push the boundaries of nanotechnology, this book invites you to join us in unraveling the mysteries of mechanical behavior at the nanoscale and unlocking new possibilities for technological advancement.

Antara Raut | Rasika Gupta | Diya Pandit

Contents

	Preface	v
1.	**Formation of Nanostructures on the Solid Surface**	**1**
	1. Introduction– 2. The electrocodeposition of nanoparticles in a metallic coating– 3.Magnetron deposition of nanofilms on the surface of a solid– 4. Formation of nanolayers on a solid surface using epitaxy– 5. Conclusions– References	
2.	**Mode-I and Mode-II Crack Tip Fields in Implicit Gradient Elasticity Based on Laplacians of Stress and Strain. Part I: Governing Equations**	**13**
	1. Introduction– 2. Preliminaries: notation– 4. Mode-I and mode-II crack problems– 5. Concluding remarks– References	
3.	**Mode-I and Mode-II Crack Tip Fields in Implicit Gradient Elasticity Based on Laplacians of Stress and Strain. Part II: Asymptotic Solutions**	**33**
	1. Introduction– 2. Summary of the governing equations for plane strain problems– 3. Near-tip asymptotic solutions for Mode-I and Mode-II loading conditions– 4. Alternative approach for the determination of the near-tip fields– 5. Discussion of the asymptotic solutions– 6. Concluding remarks– References	
4.	**Mode-I and Mode-II Crack Tip Fields in Implicit Gradient Elasticity Based on Laplacians of Stress and Strain. Part III: Numerical Simulations**	**67**
	1. Introduction– 2. Finite element formulation– 3. Square section with a circular hole– 4. Finite element analysis of crack problems– 5. Concluding remarks– References	

5. Synthesis of WO3 Nanostructures and Their Nanocomposites with Graphene Derivatives via Novel Chemical Approach **91**

1. Introduction– 2.WO3 nano-sized structures– 3.GO/WO3 nanocomposites– 4. Conclusion– References

6. Ultrasonic and Spectroscopic Techniques for the Measurement of the Elastic Properties of Nanoscale Materials **105**

1. Introduction– 2. Acoustic excitations in confined media– 3. Measurement techniques based on testing structures– 4. Measurement techniques based on ultrasonic waves– 5. Conclusions– References

Index **135**

Chapter 1

Formation of Nanostructures on the Solid Surface

Abstract

Forming nanostructures on the solids surface is one of the promising nanotechnological processes. It has been established that changes in the atomic structure of the solid surface due to the nanostructures formation result both in a significant change in various physical properties of the surface, and in an increase in its durability, strength, hardness, wear resistance. There are many different methods for forming nanostructures on solid surfaces: surface modification with nano-elements (nanoparticles, fullerenes and fullerites, graphene and nanotubes), formation of a nanocomposite layer on the surface, forming quantum dots and whiskers on the surface, implanting ions into the solid surface, laser surface treatment and other processes. The above processes are very complex and for their optimization require detailed research both by experimental and theoretical methods of mathematical modeling. The aim of this chapter was to provide a comparative review of different methods of forming nanostructures on the solids surface and mathematical modeling of these processes various aspects.

Keywords: solid surface, nanostructures, formation, physical processes, modeling

1. Introduction

Solids surface modification with the aim of forming nanostructures is widely used now. This is due to the fact that the creation of a nanostructure on the surface of a solid significantly changes its physical properties, which makes it possible to form various functional nanostructures, and increases the durability and strength of both the surface and the solid as a whole. Let us list the various technological methods of nanomodification of a solid surface.

1. Formation of a nanocomposite layer

2. Formation of quantum dots on the surface of a solid.

3. Formation of whiskers on the surface of a solid.

4. Formation of nanostructures on a porous surface

5. Surface modification by fullerenes and fullerites

6. Implantation of ions or atoms into the surface

7. Laser surface treatment

8. Formation of nanostructured coatings for spintronics

9. Healing of defects on the surface

10. Applying graphene to the surface

The list of technological processes for the modification of a solid body can be continued. The emergence of new processes at the moment is very intense.

The above technological processes can be divided into the following three main methods of nano-structural surface modification.

1. Application of a special coating having a nanostructure or including nano-structured elements in its composition.

2. Formation of nanostructured elements on the surface.

3. Changes in the atomic structure of the surface of a solid.

There are also complex methods that combine the above methods for modifying the surface of a solid.

The above processes are very complex and for their optimization require detailed research both by experimental and theoretical methods of mathematical modeling.

Experimental data show that the parameters of nanomodification processes of solids surface depend largely on the chemical, physical, and structural properties of the elements in their composition. It should also be emphasized that nanotechnological processes are complex and diverse, and their implementation typical scale is small, less than 100 nanometers. This requires a deep understanding of physical and chemical processes at nanoscale, but the nanometer scale of processes makes it difficult to study them by experimental methods only and brings about the need to use mathematical modeling. Mathematical modeling is a powerful tool, especially in new, "pioneer" fields of science and technology, in which operational experience has not yet been accumulated. Therefore, developing nanotechnology of solid surface modifications has required significant improvement and creating new methods of mathematical modeling and mathematical physics and their extension to the study of multilevel systems.

To model the class of problems considered in this review, a wide range of methods used in the simulation of meso- and nanoscale systems are used: quantum chemistry, molecular dynamics, mesodynamics and continuum mechanics. Each of these methods has its own challenges, capabilities, and limitations.

In the review, it is difficult to give in detail all the mathematical equations describing these complex processes. Therefore, the interested reader can study the exact mathematical formulation of problems in specific works given in the review or in the author's book summarizing methods for modeling nanosystems [1].

The aim of this chapter was to provide a comparative review of different methods of forming nanostructures on the solids surface and mathematical modeling of these processes various aspects.

The review presents works on three processes of solid surface modification: the electrocodeposition of nanoparticles in a metallic coating; magnetron deposition of nanofilms and the formation of nanolayers on a solid surface by epitaxy in which the author and his colleagues were directly involved.

2. The electrocodeposition of nanoparticles in a metallic coating

2.1 Technology of electrocodeposition of nanoparticles in a metallic coating

The electrocodeposition of nanoparticles in a metallic coating (EPD) process is the advanced method for using in practice the nanotechnology [2–5]. The composite coating with improved and unique operational characteristics, such as wear resistance, cracking resistance, antifriction properties, corrosion resistance, radiation resistance, and high adhesion to the substrate can be produced by this technology. Metal matrix composite electrochemical coatings (MMEC) are prepared from the suspensions, representing electrolyte solutions with additives of certain quantity of a superfine powder (**Figure 1**). The particles are adsorbed onto cathode surface in combination with metal ions during electrocodeposition (ECD) process and the metal matrix composite coating is formed. MMEC consists of galvanic metal (dispersion phase) and particles (dispersed phase).

There are the following steps of the ECD process:

1. The particles in suspension obtain a surface charge.

2. The charged particles and metal ions are transported through the liquid by the application of an electric field (electrophoresis), convection, and diffusion.

3. The particles and metal ions are adsorbed onto the electrode surface.

4. The particles adhere to the electrode surface through van der Waals forces, chemical bonding, or other forces and, simultaneously, adsorbed metal ions are reduced to metal atoms. Metal matrix is encompassed the adsorbed particles and thus the MMEC is formed.

Depending on the parameters of the co-electrochemical deposition process, strength, wear resistance, corrosion resistance, microhardness can be significantly improved. The EPD process and, consequently, the structure, morphology and properties of the composite coating are influenced by such electrochemical parameters as electrolysis conditions (chemical composition and method of stirring the electrochemical bath, presence of inclusions, temperature, pH), parameters of the applied voltage (current density, constant, pulse current), properties of inclusions (chemical composition, size, shape, surface charge, surface functionalization, concentration and dispersion of particles in an electrochemical bath), interaction between particles and electrolyte ions, nature and speed of fluid movement.

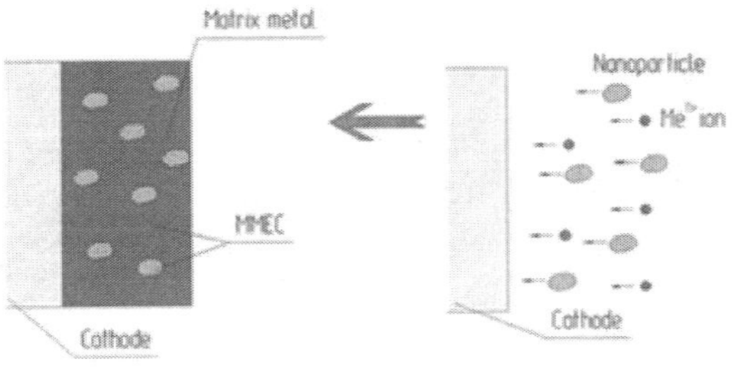

Figure 1.
The ECD process.

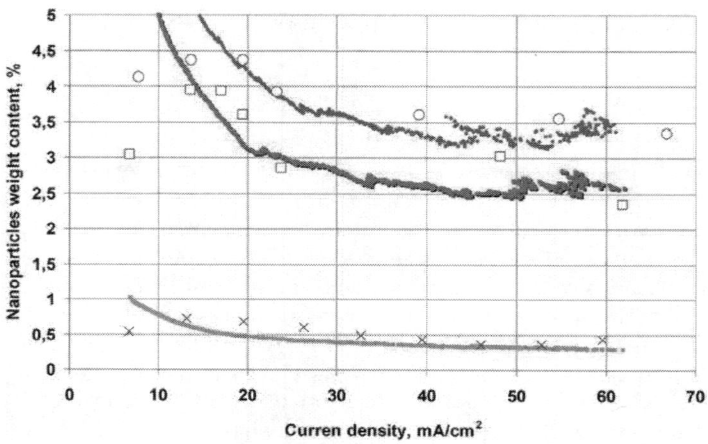

Figure 2.
Dependencies of weight content on applied current density.

The surface charge of the particles is a very important factor in the process. It should be noted that negatively charged particles are deposited more intensively, this is explained by the presence of an electric double layer around the nanoparticles. The intensity of electrolyte stirring during SEA is an important factor that affects the uniformity of the nanoparticle's distribution in the electrolyte volume due to convective flows and the delivery of particles to the cathode surface.

A large number of factors affecting the quality of the nanocomposite layer requires the use of mathematical modeling to determine the optimal parameters of the technological process of their formation.

2.2 Modeling results of nanoparticles electrocodeposition in a metallic coating

As an example, the results of mathematical modeling of copper and alumina particles ECD on RCE with consideration of electrolyte turbulent flow are depicted in **Figure 2**. A good correlation with the published experimental date [2] has been found. For the first time ever, it is found that near the RCE surface the unsteady diffusion layer is formed due electrolyte turbulent flow.

The process of nanoparticles electrocodeposition in a metallic coating can be studied in detail in [3–6].

3. Magnetron deposition of nanofilms on the surface of a solid

3.1 Technology of magnetron sputtering

In the 1970s, D. Chapin patented a planar magnetron system. This design increased the lifetime of the target and made it possible to spray on rather large areas, for example, architectural glass. Subsequently, the popularity of spray deposition grew rapidly because of the need to produce thin films with a uniform composition and good adhesion to the substrate surface, the demand is due to microelectronics [7–10].

The diagram of the magnetron chamber shown in **Figure 3** illustrates the process of magnetron sputtering.

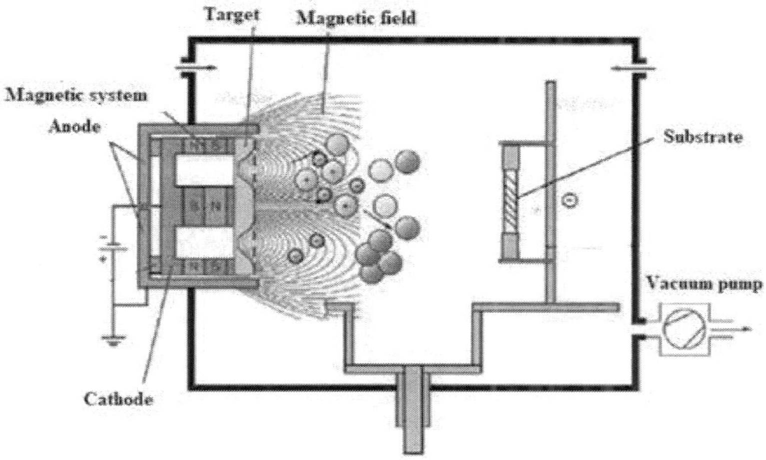

Figure 3.
Diagram of the chamber of the installation of magnetron sputtering [7].

In the above scheme, a planar target is used, colored yellow. An inert gas is required to create a plasma above the target. The black lines show the magnetic field holding the plasma; in this area, the atoms of the working gas are ionized. Then, under the influence of an electric field, the ions knock the atoms out of the target. The active gas serves to form oxide and nitride films. A bias potential is applied to the substrate, the value of which affects the structure of the formed coatings. When a negative bias potential is applied to the substrate with respect to the magnetron plasma, the ions are accelerated in the electric field of the substrate and carry out a low-energy bombardment of the substrate surface.

Magnetron sputtering is widely used for forming of new electronic devices based on new functional nanosystems, which requires a thorough study of their properties at the atomic and molecular level. Their functional characteristics depend on the structure and nanosystems morphology: optical absorption, electromagnetic parameters, etc. Recently, considerable attention has been paid to such nanosystems as superconductors, magnetics, heterostructures for spintronics. These systems are layered or nano-dispersed composites with local inhomogeneities zones. The local inhomogeneities presence results in forming stable vortex-like regions of reverse magnetization, called magnetic skyrmions and anti-skyrmions, which are very promising tools for use in the spintronics field.

3.2 Results of modeling magnetron sputtering of multilayer nanosystems for spintronics

A particularly important effect on the spintronic objects and other nanoscale elements properties is exerted by the structure of layers, domains, regions, films arising during their manufacture [11–13]. The structure generates at the atomic level. Therefore, the main mechanisms of regulation, activation, and interaction in spin nanosystems must be monitored at this scale. Atomistic modeling makes it possible to identify the modes and technological processes parameters when the functional nanostructure's structure is the closest to the required characteristics. There is an acute problem of forming multilayer nanosystem with clear boundaries of various nanolayers sections.

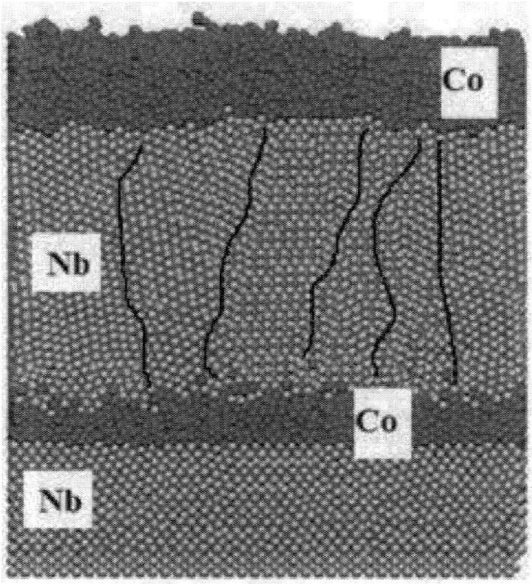

Figure 4.
Nb-Co multilayer quantum valve.

As an example illustrating the simulation of the real structure of a material, let us consider the structure of a multilayer nanosystem of an Nb-Co spin valve.

Figure 4 well characterize the qualitative picture of a spin valve forming from niobium and cobalt layers and the structure of the layers. The structure of the layers formed by niobium atoms is close to crystalline. In this case, groups of atoms are combined into nanocrystallites with vertical spatial orientation. In **Figure 4**, solid lines mark the boundaries of these nanocrystals. Cobalt nanofilms have an amorphous structure. The results obtained are in good agreement with experimental studies of the structure of various multilayer nanosystems [14].

4. Formation of nanolayers on a solid surface using epitaxy

4.1 Technology for the formation of nanolayers on a solid surface by epitaxy

At present, epitaxy is widely used to modify the surface of solids. Epitaxy is the process of building up layers on the surface of a solid [15]. **Figure 5** shows a simplified diagram of the process.

The structure of the formed epitaxial layers, as a rule, reproduces the structure of the solid's surface, and the chemical composition of the epitaxial layer and the substrate may differ. In the process of epitaxy, the chemical composition of the grown layers can be controlled both gradually and discretely. This technology makes it possible to grow multilayer nanosystems with a thickness up to atomic dimensions. The structures grown in this way are nanostructures: nanolayers, quantum dots, whiskers, etc. give the surface of a solid body unique physical properties that are absent in the base material. There are different types of epitaxies. If the materials of the resulting layer and the substrate are the same, then the process is called auto epitaxial or homoepitaxial. If the materials of the layer and the substrate are different, then the process is called heteroepitaxial.

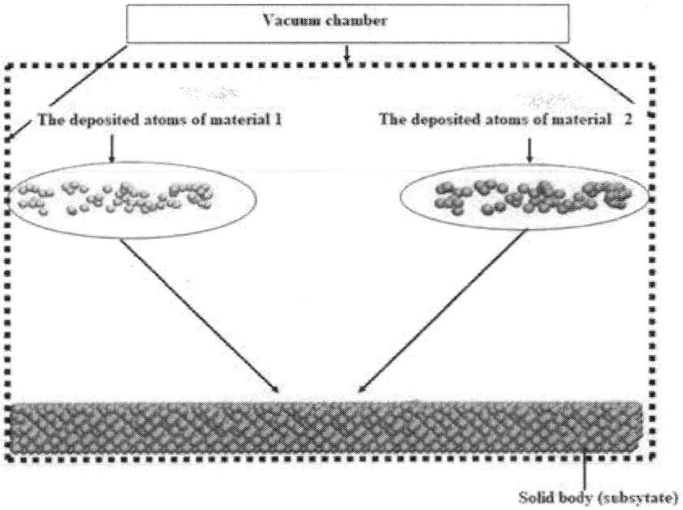

Figure 5.
Epitaxy scheme.

4.2 Results of modeling the formation of nanostructures on a solid surface by epitaxy

4.2.1 Simulation of the formation of quantum dots

As examples of modeling process for the formation of quantum dots on the a solid surface, let us consider the formation of gold nanoparticles on the silicon surface and the formation of a complex system consisting of gallium and antimony atoms. The simulation results are presented in **Figures 6** and 7, respectively.

For the simulation, a silicon substrate with orientation (100) was used, which is indicated in green in **Figures 6** and 7. Gold atoms, indicated in yellow in the figure, were deposited on the substrate. Then, silicon atoms were deposited on the resulting system. The temperature of the simulated system was kept constant and equal to 800 K.

As can be seen from **Figure 6**, gold nanoparticles of various diameters are formed on the substrate. The physical process of forming the gold-silicon nanosystem is interesting. First, gold atoms are deposited on a silicon substrate, and then they are collected in nanoparticles (quantum dots) of various sizes, spherical.

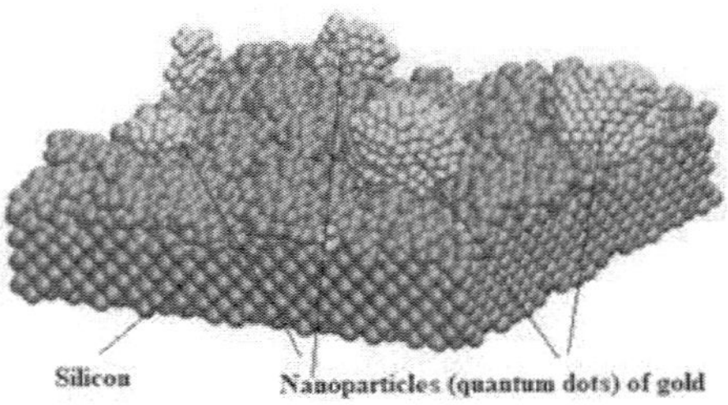

Figure 6.
Picture of a nanosystem on a solid surface obtained by simulating the deposition of gold atoms on a silicon substrate with the orientation (100).

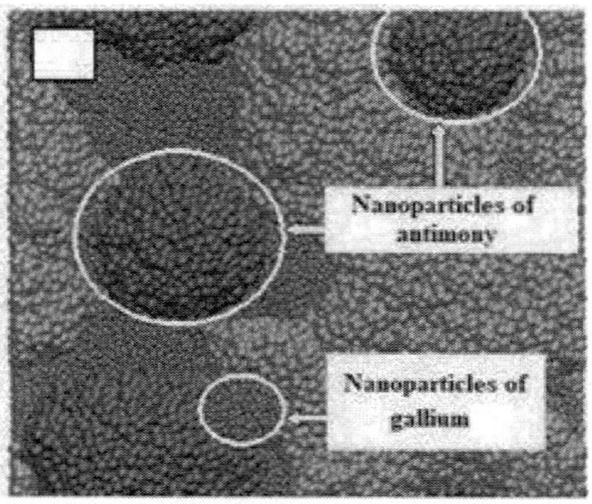

Figure 7.
Pictures of a nanosystem on a solid surface obtained by simulating the deposition of gallium and antimony atoms on a silicon substrate with the orientation (100).

As a result of modeling the formation of a complex system consisting of gallium (dark green) and antimony (lilac) atoms, gallium, and antimony atoms form conglomerates of various shapes) (**Figure 7**). Diffusion of gallium and antimony atoms into the substrate is not observed. It can be seen from the presented figures that the atoms of the substances under consideration tend to form nanoparticles on the silicon surface of various shapes. Silicon atoms fill the entire space between the formed agglomerates of gallium and antimony nanoparticles.

4.2.2 Modeling the formation of nanoelements on a porous solid surface

Let us consider the results of modeling the processes of forming ZnS nanolayers on the surface of porous aluminum oxide on matrices. Such nanostructures are actively used in optical systems in the infrared range [16].

The process of nanostructure formation in zinc sulfide nanofilms is illustrated in **Figure 8**.

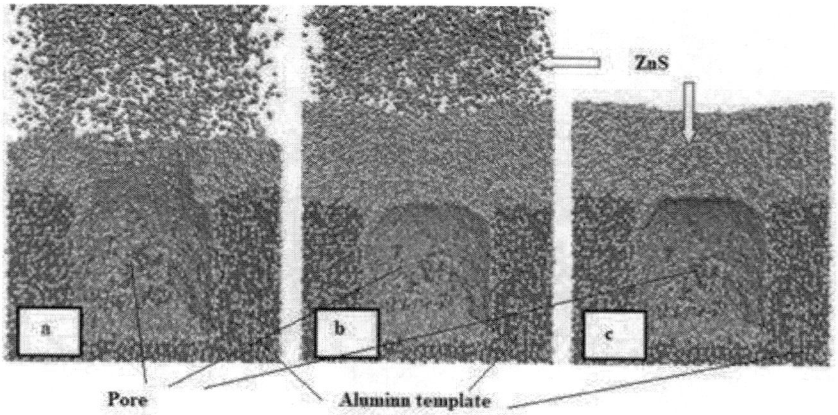

Figure 8.
Pictures of the sequential formation of a ZnS nanolayer on a porous alumina substrate for the deposition time: (a) 0.2 ns, (b) 0.4 ns, and (c) 0.6 ns.

Analysis of the nanostructure shows that the overgrowth of pores on the surface of a solid with the indicated atoms occurs gradually. First, a ZnS nanolayer begins to form near the pore (**Figure 8a**), which subsequently gradually closes the entire pore. Zinc sulfide molecules partially enter the pore, but its complete dense filling does not occur (**Figure 8b** and **c**). Nevertheless, by the end of deposition, almost the entire inner surface of the pore is covered with ZnS molecules. The gradual filling of the pore leads to the appearance of round growths over the pore.

In general, the surface of the ZnS nanolayer forms even, with a slight decrease in the thickness of the nanolayer above the pore area. The formation of molecular agglomerates in the space above the substrate during epitaxy is not observed; therefore, the nanolayer is even and there are no significant differences in the surface relief. The growth rate of the nanofilm during the deposition process was uniform. The thickness of the formed nanolayer was 6.6–6.8 nm.

Analysis of the materials structure based on zinc sulfide indicates a predominantly amorphous structure of templates and formed nanofilms with insignificant crystallization areas with different spatial orientations.

5. Conclusions

The chapter provides methods for modifying the surface of a solid and gives examples of modeling three modification processes. It is shown that mathematical modeling makes it possible to predict the structure of the modified surface and determine the parameters of technological processes of modification. To obtain more complete information on modeling methods, the reader can obtain additional information from the literature on modification methods [17–19] and modeling methods [20–24].

References

[1] A. V. Vakhrushev Computational Multiscale Modeling of Multiphase Nanosystems. Theory and Applications, Apple Academic Press, Waretown, New Jersey, 2017. 402 p. DOI: 10.1088/0965-0393/14/6/007

[2] Stojak, J. L.; Fransaer, J.; Talbot, J. B. Review of Electrocodeposition. In Advances in Electrochemical Science and Engineering; Alkire, R. C., Kolb, D. M., Eds.; Wiley-VCH Verlag: Weinheim, 2002; vol. 9, p 457.

[3] Bicelli, L. P.; Bozzini, B.; Mele, C. A Review of Nanostructural Aspects of Metal Electrodeposition. J. Electrochem. Soc. 2008, 3, 356-408.

[4] Hovestad, A.; Janssen, L. J. J. Electrochemical Codeposition of Inert Particles in a Metallic Matrix. J. Appl. Electrochem. 1995, 25, 519-527.

[5] Gomes, A.; Pereira, I.; Fernández, B.; Pereir, R. Electrodeposition of Metal Matrix Nanocomposites: Improvement of the Chemical Characterization Techniques. In Advances in Nanocomposites—Synthesis, Characterization and Industrial Applications; Boreddy, R., Ed.; InTech: Rijeka, 2011; p. 538.

[6] Vakhrushev, A. V.; Molchanov, E. K. Hydrodynamic Modeling of Electrocodeposition on a Rotating Cylinder Electrode. Key Eng. Mater. 2015, 654, 29-33.

[7] Brauer, G. Magnetron sputtering – Milestones of 30 years / G. Brauer, B. Szyszka, M. Vergohl, R. Bandorf. // Vacuum. – 2010. – Vol. 84. – P. 1354-1359.

[8] Waits, R.K. Planar magnetron sputtering / R. K. Waits // Journal of Vacuum Science and Technology. – 1978. – Vol. 15. – P. 179-187.

[9] Kelly, P.J. Magnetron sputtering: a review of recent developments and applications / P.J. Kelly and R.D. Arnell. // Vacuum. – 2000. – Vol. 56. – P. 159-172.

[10] Anders, A. Tutorial: Reactive high-power impulse magnetron sputtering (R-HiPIMS) / A. Anders // Journal of Applied Physics. – 2017. – Vol. 121. – P. 171101-1 – 171101-34.

[11] D. Lenk, R. Morari, V.I. Zdravkov, A. Ullrich, Y. Khaydukov, G. Obermeier, C. Mueller, A.S. Sidorenko, K. H.-A. von Nidda, S. Horn, L.R. Tagirov, R. Tidecks, Full Switching FSF-type Superconducting Spin-Triplet MRAM-Element, R. *Phys. Rev. B.* 96 (2017) 184521/1-184521/18.

[12] L. Lazar, K. Westerholt, H. Zabel, L. R. Tagirov, Yu. V. Goryunov, N. N Garifiyanov, I. A. Garifullin, Superconductor/ferromagnet proximity effect in Fe/Pb/Fe trilayers, *Phys. Rev. B.* 61 (2000) 3711-3722

[13] N. Klenov, Y. Khaydukov, S. Bakurskiy, R. Morari, I. Soloviev, V. Boian, T. Keller, M. Kupriyanov, A. Sidorenko, B. Keimer, Periodic Co/Nb pseudo spin valve for cryogenic memory, Beilstein J. Nanotechnol. 10 (2019) 833-839.

[14] A.V. Vakhrushev, A. Yu. Fedotov, V. Boian, R. Morari, A.S. Sidorenko, Molecular dynamics modeling of formation processes parameters influence on a superconducting spin valve structure and morphology, Beilstein Archives. 202067 (2020), 26p. doi: 10.3762/bxiv.2020.67.v1

[15] McCray, W.P. "MBE Deserves a Place in the History Books". Nature Nanotechnology. 2 (5): (2007). 259-261

[16] R.G. Valeev, A.V. Vakhrushev, A. Yu. Fedotov, D.I. Petukhov, Functional

Semiconductor Nanostructures in Porous Anodic Alumina Matrices: Modeling, Synthesis, Properties, Apple Academic Press, Waretown, 2019.

[17] Suvorov S.V., Severyukhin A.V., Vakhrushev A.V. The effect of the angle of «meeting» of fullerite C60 with a solid substrate on the deposition process. PNRPU Mechanics Bulletin, 2020, no. 3, pp. 90-97. DOI: 10.15593/perm.mech/2020.3.09

[18] A.A. Shushkov, P.V. Bykov V.L. Vorob'ev , A.V. Vakhrushev, V. Ya. Bayankin Research of physical and mechanical properties of surface layers of multilayer ti/al films after ion-beam mixing. Chemical Physics and Mesoscopy, 2021, vol. 23, no. 2, pp. 196-211. DOI: 10.15350/17270529.2021.2.18

[19] V. L. Vorob'ev, I. N. Klimova, A. A. Kolotov, P. V. Bykov & V. Ya. Bayankin Formation of a Nickel–Aluminum Intermetallic during Ion-Beam Mixing / Journal of Surface Investigation: X-ray, Synchrotron and Neutron Techniques volume 13, pages 1225-1229 (2019)

[20] A.V. Vakhrushev (Ed.) Molecular Dynamics. Intech, London. UK. 2018. - 104 p

[21] Martin Steinhauser O. Computational Multiscale Modelling of Fluids and Solids. Theory and Application. Berlin–Heidelberg: Springer-Verlag; 2008. 427p

[22] E. Weinan Principles of Multiscale Modeling. Cambridge: Cambridge University Press; 2011. 466p

[23] Mercle CR. Computational nanotechnology. Nanotechnology. 1991;2:134-141

[24] Marx D, Hutter J. Ab Initio Molecular Dynamics: Theory and Advanced Methods. New York: Cambridge University Press; 2009. 567p

Chapter 2

Mode-I and Mode-II Crack Tip Fields in Implicit Gradient Elasticity Based on Laplacians of Stress and Strain. Part I: Governing Equations

Abstract

Models of implicit gradient elasticity based on Laplacians of stress and strain can be established in analogy to the models of linear viscoelastic solids. The most simple implicit gradient elasticity model including both, the Laplacian of stress and the Laplacian of strain, is the counterpart of the three-parameter viscoelastic solid. The main investigations in Parts *I*, *II*, and *III* concern the "three-parameter gradient elasticity model" and focus on the near-tip fields of Mode-I and Mode-II crack problems. It is proved that, for the boundary and symmetry conditions assumed in the present work, the model does not avoid the well-known singularities of classical elasticity. Nevertheless, there are significant differences in the form of the asymptotic solutions in comparison to the classical elasticity. These differences are discussed in detail on the basis of closed-form analytical solutions. Part *I* provides the governing equations and the required boundary and symmetry conditions for the considered crack problems.

Keywords: implicit gradient elasticity, Laplacians of stress, Laplacians of strain, micromorphic and micro-strain elasticity, plane strain state

1. Introduction

The most simple constitutive law in explicit gradient elasticity is the model with equation:

$$\Sigma_{ij} = \mathbb{C}_{ijmn}\,\varepsilon_{mn} - c_2\,\mathbb{C}_{ijmn}\,(\Delta\varepsilon)_{mn} \quad (\mathrm{KG-Model}). \tag{1}$$

Here, $\boldsymbol{\Sigma} = \boldsymbol{\Sigma}^T$ is the Cauchy stress tensor, ε is the strain tensor, $\mathbb{C}$ is the isotropic elasticity tensor, Δ is the Laplacian operator, and c_2 is a material parameter, with $\sqrt{c_2}$ denoting an internal material length. The components in Eq. (1) are referred to a Cartesian coordinate system. It seems that the constitutive law (1) has been introduced for the first time by Altan and Aifantis [1]. These authors (cf. also Georgiadis [2]) showed that the constitutive Eq. (1) leads to regular strain solutions at the crack tip of Mode-III crack problems. However, the stress field remains singular at the crack tip as in the case of classical elasticity. Moreover, Altan and Aifantis [1], as well as Georgiadis [2], presented an appropriate isotropic energy function for the mechanical model in Eq. (1) in the context of Mindlins gradient elasticity theory (see Mindlin [3] as well as Mindlin and Eshel [4]). An alternative approach to this model has been proposed in Broese et al. [5], where an analogy between gradient elasticity models and linear viscoelastic solids is established. According to this analogy, Eq. (1) is regarded as the gradient elasticity counterpart of the Kelvin viscoelastic solid. The short hand notation "KG-Model" in Eq. (1) stands for "Kelvin-Gradient-Elasticity-Model."

Now, the question arises, if a gradient elasticity model including both, the Laplacian of stress and the Laplacian of strain, could remove both, the singularities of stress and the singularities of strain at the crack tip (cf. Gutkin and Aifantis [6]). The most simple generalization of Eq. (1), including the Laplacians of stress and strain, reads as follows:

$$\Sigma_{ij} - c_1 (\Delta \mathbf{\Sigma})_{ij} = \mathbb{C}_{ijmn}\, \varepsilon_{mn} - c_2 \mathbb{C}_{ijmn} (\Delta \varepsilon)_{mn} \quad (3 - \text{PG} - \text{Model}), \qquad (2)$$

where the same notation as in Eq. (1) applies and $\sqrt{c_1}$ is a further internal material length. To our knowledge, model (2) has been introduced for the first time by Gutkin and Aifantis [6]. These authors proposed the gradient elasticity law (2) ad hoc in an attempt to eliminate the singularities of stress and strain of defects. Equations of the form (2) are known as models of implicit gradient elasticity (see Askes and Gutiérrez [7]).

Broese et al. [5] proved that Eq. (2) can be derived as a particular case of Mindlins micro-structured elasticity, which arises whenever the micro-deformation of the micromorphic continuum is supposed to be a symmetric tensor. Because the micro-structured elastic continuum of Mindlin and the micromorphic elastic continuum of Eringen (see, e.g., Eringen and Suhubi [8] and Eringen [9]) are essentially equivalent to each other, in the present work we will call both as micromorphic continua. According to Forest and Sievert [10], the resulting micromorphic theory is named micro-strain theory. It is shown in Broese et al. [5] that, in the context of micro-strain elasticity, the 3-PG-Model (2) can be derived as a combination of elasticity constitutive laws and the equilibrium equation for the so-called double stress.

On the other hand, Broese et al. [5] showed that Eq. (2) can be established alternatively by supposing the continuum to be classical, i.e., exhibiting only classical displacement degrees of freedom, but in the framework of the non-conventional thermodynamics proposed in Alber et al. [11]. To be more specific, the micro-deformation variable of the micro-strain approach has to be viewed as an internal state variable analogous to the inelastic strain in linear viscoelasticity. Eq. (2) then turns out to be a constitutive law, which is the counterpart in gradient elasticity of the three-parameter viscoelastic solid. The short hand notation "3-PG-Model" stands for "3-Parameter-Gradient-Elasticity-Model." A general analogy to the constitutive laws describing viscoelastic solids can be established by using a nonstandard spring in gradient elasticity corresponding to the dashpot element in linear viscoelasticity and the Laplacian operator Δ in place of the ordinary time derivative in the evolution laws of dashpot elements. Using virtual power balance arguments, Broese et al. [5] derived the same boundary conditions along the lines of the second approach as in the micro-strain approach. But now the boundary conditions have to be understood as constitutive boundary conditions, analogous to the constitutive initial conditions in viscoelasticity.

Because all resulting governing equations and boundary conditions in the two approaches are equal to each other, we shall proceed further by regarding the

3-PG-Model as a particular case of the micro-strain elasticity. The present work (Parts *I*, *II*, and *III*) is concerned with the near-tip fields predicted by the 3-PG-Model for Mode-I and Mode-II types of crack problems. Unlike statements made somewhere else (see Part *II*), we prove that, for the assumptions made here, the 3-PG-Model does not eliminate the well-known singularities of classical elasticity. Nevertheless, compared with the form of asymptotic solutions in classical elasticity, there are interesting new aspects, which are discussed in detail in Part *II* on the basis of closed-form analytical solutions. Part *I* provides the governing equations and the required boundary and symmetry conditions in order to establish the analytical solutions.

2. Preliminaries: notation

Throughout the paper, we largely use the same notation as in Mindlin [3] and Mindlin and Eshel [4], in order to facilitate the comparison with these works. The deformations are assumed to be small, so we do not distinguish, as usually done, between reference and actual configuration. All indices will have the range of integers $(1, 2, 3)$, while summation over repeated indices is implied. Explicit reference to space and time variables, upon which a function may depend, will be dropped in most part of the paper. Also, we shall not distinguish between functions and their values. However, if necessary, we shall give explicitly the set of variables which the function depends on.

Let $\mathcal{B}$ be a material body which may be identified by the position vectors $\boldsymbol{x} = x_i \boldsymbol{e}_i$, with respect to a Cartesian coordinate system $\{x_i\}$ inducing the orthonormal basis $\{\boldsymbol{e}_i\}$. The body $\mathcal{B}$ occupies the space V in the three-dimensional Euclidean space we deal with. We indicate by $\boldsymbol{n}$ the outward unit normal vector to the surface ∂V bounding the space V. Small Latin indices will be used in conjunction with Cartesian coordinates and related components. If f is a function of the Cartesian coordinates x_i, then we shall use the notations for partial derivatives as follows:

$$\partial_i f := \frac{\partial f}{\partial x_i}, \quad \partial_{ij} f := \frac{\partial^2 f}{\partial x_i \partial x_j}. \tag{3}$$

Let $\nabla = \partial_i \boldsymbol{e}_i$ be the nabla operator and $\mathfrak{a}$ be some vector or higher order tensor. The gradient, the divergence, and the Laplacian of $\mathfrak{a}$ are defined, respectively, by $\operatorname{grad} \mathfrak{a} \equiv \nabla \mathfrak{a} := \nabla \otimes \mathfrak{a}$, $\operatorname{div} \mathfrak{a} := \nabla \cdot \mathfrak{a}$, and $\Delta \mathfrak{a} = \operatorname{div} \operatorname{grad} \mathfrak{a}$, where $\cdot$ and $\otimes$ are the scalar and the tensorial products between two vectors. It is helpful to use notations for components of the form $(\mathfrak{a})_{ij\ldots} = \mathfrak{a}_{ij\ldots}$. Thus, if A_{ij} and $\mathcal{A}_{ijk}$ are the Cartesian components of a second-order tensor $\boldsymbol{A}$ and a third-order tensor $\boldsymbol{\mathcal{A}}$, then, with respect to the Cartesian basis $\{\boldsymbol{e}_i\}$, we have the following equations:

$$(\nabla \boldsymbol{A})_{ijk} = \partial_i A_{jk}, \tag{4}$$

$$(\operatorname{div} \boldsymbol{A})_i = \partial_j A_{ji}, \tag{5}$$

$$(\Delta \boldsymbol{A})_{ij} = \partial_{kk} A_{ij}, \tag{6}$$

$$(\operatorname{div} \boldsymbol{\mathcal{A}})_{ij} = \partial_k \mathcal{A}_{kij}. \tag{7}$$

We denote by $\mathbb{C}$ the fourth-order isotropic elasticity tensor with Cartesian components as follows:

$$\mathbb{C}_{ijmn} = \lambda \delta_{ij} \delta_{mn} + \mu \left(\delta_{im} \delta_{jn} + \delta_{in} \delta_{jm} \right), \tag{8}$$

where λ and μ are the Lamé constants and δ_{ij} is the Kronecker delta. For some calculations, it will be convenient to use the Young's modulus E and the Poisson ratio ν,

$$\nu = \frac{\lambda}{2(\lambda+\mu)}, \quad E = 2\mu(1+\nu). \tag{9}$$

Since $\mathbb{C}$ satisfies the symmetry properties

$$\mathbb{C}_{ijmn} = \mathbb{C}_{jimn} = \mathbb{C}_{mnij} = \mathbb{C}_{ijnm}, \tag{10}$$

we have for every second-order tensor $\boldsymbol{A}$, that

$$\mathbb{C}_{ijmn} A_{mn} = A_{mn} \mathbb{C}_{mnij}. \tag{11}$$

For any tensor $\mathfrak{a}$ with components $\mathfrak{a}_{i\,\ldots jk\ldots p}$, we write $\mathfrak{a}_{i\,\ldots(jk)\ldots p}$ for its symmetric part with respect to the indices j and k. Thus, if $\boldsymbol{A}^{(s)}$ is the symmetric part of a second-order tensor $\boldsymbol{A}$, then $A_{(ij)} = A^{(s)}_{ij}$. Corresponding notations apply with regard to components related to curvilinear coordinate systems. Of particular interest for our work are cylindrical coordinates $\{r, \varphi, z\}$. We find it convenient to use Greek indices $\alpha, \beta, \ldots$ to indicate both, physical components with respect to cylindrical coordinates and cylindrical coordinates itself. Thus, e.g., we write $A_{\alpha\beta}$ for the physical components of the second-order tensor $\boldsymbol{A}$ and denote by $[A_{\alpha\beta}]$ the matrix of components,

$$[A_{\alpha\beta}] = \begin{pmatrix} A_{rr} & A_{r\varphi} & A_{rz} \\ A_{\varphi r} & A_{\varphi\varphi} & A_{\varphi z} \\ A_{zr} & A_{z\varphi} & A_{zz} \end{pmatrix}. \tag{12}$$

Similar notations hold for any tensor of arbitrary order. The summation convention applies in analogous manner, e.g., we have $A_{\alpha\alpha} = A_{rr} + A_{\varphi\varphi} + A_{zz}$. Because cylindrical coordinate systems are orthogonal, the algebraic operations between the corresponding physical components of tensors are identical in form to those with respect to Cartesian components. Moreover, the physical components of isotropic tensors are identical to their Cartesian components, e.g., the physical components of the isotropic elasticity tensor $\mathbb{C}$ in Eq. (8) are given as follows:

$$\mathbb{C}_{\alpha\beta\gamma\zeta} = \lambda\delta_{\alpha\beta}\delta_{\gamma\zeta} + \mu\left(\delta_{\alpha\gamma}\delta_{\beta\zeta} + \delta_{\alpha\zeta}\delta_{\beta\gamma}\right). \tag{13}$$

The physical components with respect to cylindrical coordinates of $\nabla\boldsymbol{A}$, $\Delta\boldsymbol{A}$ and $\operatorname{div}\boldsymbol{A}$ are calculated in A. For partial derivatives of a function f with respect to cylindrical coordinates, we use notations, in analogy to Eq. (3), of the forms

$$\partial_r f := \frac{\partial f}{\partial r}, \quad \partial_\varphi f := \frac{\partial f}{\partial \varphi}, \quad \partial_z f := \frac{\partial f}{\partial z}, \tag{14}$$

$$\partial_{rr} f := \frac{\partial^2 f}{\partial r\,\partial r}, \ \ldots. \tag{15}$$

3. Governing equations for the 3-PG-Model

This section provides a short overview about the 3-PG-Model in a form which is adequate for developing analytical solutions.

3.1 The 3-PG-Model as particular case of micro-strain elasticity

Assume the material body to be a micromorphic continuum. Besides the classical kinematical degrees of freedom, micromorphic continua are characterized by additional degrees of freedom due to the deformations of the micro-continua, which are assumed to be attached at every point of the macro-continuum (see Mindlin [3] and Broese et al. [5]). Therefore, in the micromorphic continuum theory, a nonclassical (double) stress and a nonclassical stress power are introduced in addition to the classical ones, but otherwise the theory is formulated in the framework of classical thermodynamics.

Let $\boldsymbol{\Psi}$ be the micro-deformation tensor of a micromorphic continuum, $\boldsymbol{u}$ be the macro-displacement vector, and ε be the macro-strain tensor,

$$\varepsilon_{ij} \equiv \varepsilon_{(ij)} := \frac{1}{2}\left(\partial_i u_j + \partial_j u_i\right). \tag{16}$$

All component representations in Section 3 are referred to the Cartesian coordinate system $\{x_i\}$. Assume $\boldsymbol{\Psi}$ and the so-called relative deformation γ to be symmetric,

$$\Psi_{ij} \equiv \Psi_{(ij)}, \quad \gamma_{ij} \equiv \gamma_{(ij)} := \varepsilon_{ij} - \Psi_{ij}. \tag{17}$$

This means that $\boldsymbol{\Psi}$ and γ are strain tensors and that the components of the so-called micro-deformation gradient $\boldsymbol{k}$,

$$k_{ijk} := (\nabla\boldsymbol{\Psi})_{ijk} = \partial_i \Psi_{jk}, \tag{18}$$

exhibit the symmetry property

$$k_{ijk} \equiv k_{i(jk)}. \tag{19}$$

Following Forest and Sievert [10], we denote a micromorphic elasticity theory based on Eqs. (16)–(18) as micro-strain elasticity.

According to Broese et al. [5], the 3-PG-model can be established as a particular case of the micro-strain elasticity by assuming the existence of a free energy (per unit macro-volume) ψ of the form:

$$\begin{aligned}\psi = \psi(\varepsilon, \gamma, \boldsymbol{k}) =& \frac{1}{2}\,\varepsilon_{ij}\,\mathbb{C}_{ijmn}\,\varepsilon_{mn} + \frac{1}{2}\frac{c_2 - c_1}{c_1}\,\gamma_{ij}\,\mathbb{C}_{ijmn}\,\gamma_{mn} \\ &+ \frac{1}{2}\,(c_2 - c_1)\,k_{ijk}\,\mathbb{C}_{jkmn}\,k_{imn}.\end{aligned} \tag{20}$$

The components $\mathbb{C}_{ijmn}$ are defined in Eq. (8) and c_1 as well as c_2 are scalar parameters constrained to $c_2 > c_1 > 0$ with $\sqrt{c_1}$ and $\sqrt{c_2}$ denoting internal material lengths as noted in Section 1. The Cauchy stress tensor $\boldsymbol{\Sigma}$ is then given by (cf. Broese et al. [5]):

$$\Sigma_{ij} \equiv \Sigma_{(ij)} = \tau_{ij} + \sigma_{ij} = \frac{c_2}{c_1}\,\mathbb{C}_{ijmn}\,\varepsilon_{mn} - \frac{c_2 - c_1}{c_1}\,\mathbb{C}_{ijmn}\,\Psi_{mn}, \tag{21}$$

where

$$\tau_{ij} \equiv \tau_{(ij)} = \frac{\partial\psi}{\partial\varepsilon_{ij}} = \mathbb{C}_{ijmn}\,\varepsilon_{mn}, \tag{22}$$

$$\sigma_{ij} \equiv \sigma_{(ij)} = \frac{\partial \psi}{\partial \gamma_{ij}} = \frac{c_2 - c_1}{c_1} \left(\mathbb{C}_{ijmn} \, \varepsilon_{mn} - \mathbb{C}_{ijmn} \, \Psi_{mn} \right). \tag{23}$$

Further, there exists a double stress $\boldsymbol{\mu}$ which satisfies the potential relation

$$\mu_{ijk} \equiv \mu_{i(jk)} = \frac{\partial \psi}{\partial k_{ijk}} = (c_2 - c_1) \, \partial_i \Psi_{mn} \, \mathbb{C}_{mnjk}. \tag{24}$$

For static problems, the classical and nonclassical stresses have to satisfy corresponding equilibrium equations. In the absence of body forces and body double forces, these are (see Mindlin [3] or Broese et al. [5])

$$\partial_j \Sigma_{ji} = 0, \tag{25}$$

$$\partial_k \mu_{kij} + \sigma_{ij} = 0. \tag{26}$$

The concomitant classical and nonclassical boundary conditions are as follows:

$$\text{Either } P_i = n_j \, \Sigma_{ji} \text{ or } u_i \text{ (class.bound.cond.)}, \tag{27}$$

$$\text{and either } T_{ij} = n_k \, \mu_{kij} \text{ or } \Psi_{ij} \text{ (non} - \text{class.bound.cond.)}, \tag{28}$$

have to be prescribed on the boundary ∂V.

The 3-PG-Model can be obtained from the above equations by first inserting Eq. (24) into Eq. (26), as follows:

$$\begin{aligned} \sigma_{ij} &= -\partial_k \mu_{kij} = -(c_2 - c_1) \, \partial_{kk} \Psi_{mn} \, \mathbb{C}_{mnij} \\ &= -(c_2 - c_1) \, \mathbb{C}_{ijmn} \, (\Delta \boldsymbol{\Psi})_{mn}. \end{aligned} \tag{29}$$

Then take the Laplacian of Eq. (21), as follows:

$$(\Delta \boldsymbol{\Sigma})_{ij} = \frac{c_2}{c_1} \, \mathbb{C}_{ijmn} \, (\Delta \varepsilon)_{mn} - \frac{c_2 - c_1}{c_1} \, \mathbb{C}_{ijmn} \, (\Delta \boldsymbol{\Psi})_{mn}, \tag{30}$$

and use Eq. (21) as well as Eq. (22) in Eq. (29), as follows:

$$-(c_2 - c_1) \, \mathbb{C}_{ijmn} \, (\Delta \boldsymbol{\Psi})_{mn} = \sigma_{ij} = \Sigma_{ij} - \mathbb{C}_{ijmn} \, \varepsilon_{mn}. \tag{31}$$

The latter together with Eq. (30) yield the following equation:

$$\Sigma_{ij} - c_1 \, (\Delta \boldsymbol{\Sigma})_{ij} = \mathbb{C}_{ijmn} \, \varepsilon_{mn} - c_2 \, \mathbb{C}_{ijmn} \, (\Delta \varepsilon)_{mn}, \tag{32}$$

which is nothing but the 3-PG-Model (2).

3.1.1 A useful equation for Ψ

For later reference, we derive a useful equation for the strain $\boldsymbol{\Psi}$. When seeking analytical solutions, there are two possibilities, either to find solutions in terms of $\boldsymbol{\Psi}$ or in terms of $\boldsymbol{\mu}$. In the first case, $\boldsymbol{\mu}$ is eliminated at the cost of a higher order partial differential equation, but no compatibility conditions for $\boldsymbol{\Psi}$ are needed. To be more specific, assuming $\mathbb{C}^{-1}$ to exist, we infer from Eqs. (23) and (29) that

$$\varepsilon_{ij} = \Psi_{ij} - c_1 \, (\Delta \boldsymbol{\Psi})_{ij}. \tag{33}$$

Further, from Eq. (21), we get the following equation:

$$\varepsilon_{ij} = \frac{c_1}{c_2} \left(\mathbb{C}^{-1}\right)_{ijmn} \Sigma_{mn} + \frac{c_2 - c_1}{c_2} \Psi_{ij}. \tag{34}$$

By combining the last two equations, we gain the useful relation as follows:

$$\Psi_{ij} - c_2 (\Delta \boldsymbol{\Psi})_{ij} - \left(\mathbb{C}^{-1}\right)_{ijmn} \Sigma_{mn} = 0. \tag{35}$$

For given $\boldsymbol{\Sigma}$, this is a (Helmholtz) partial differential equation for the components of $\boldsymbol{\Psi}$.

4. Mode-I and mode-II crack problems

In Part *II*, we consider Mode-I and Mode-II loading conditions for a sharp crack in the context of plane strain problems and employ cylindrical coordinates $\{r, \varphi, z\}$ as indicated in **Figure 1**. The aim of this section is to set up all relevant equations which are needed in Part *II*.

4.1 Kinematics

Plane strain state of micro-strain continua in equilibrium is characterized by the assumptions that

$$[u_\alpha] = \begin{pmatrix} u_r \\ u_\varphi \\ 0 \end{pmatrix}, \quad [\varepsilon_{\alpha\beta}] = \begin{pmatrix} \varepsilon_{rr} & \varepsilon_{r\varphi} & 0 \\ \varepsilon_{r\varphi} & \varepsilon_{\varphi\varphi} & 0 \\ 0 & 0 & 0 \end{pmatrix}, \tag{36}$$

$$[\Psi_{\alpha\beta}] = \begin{pmatrix} \Psi_{rr} & \Psi_{r\varphi} & 0 \\ \Psi_{r\varphi} & \Psi_{\varphi\varphi} & 0 \\ 0 & 0 & 0 \end{pmatrix}, \tag{37}$$

and that $\boldsymbol{u}$, ε, and $\boldsymbol{\Psi}$ are independent of z,

$$u_\alpha = u_\alpha(r, \varphi), \quad \varepsilon_{\alpha\beta} = \varepsilon_{\alpha\beta}(r, \varphi), \quad \Psi_{\alpha\beta} = \Psi_{\alpha\beta}(r, \varphi). \tag{38}$$

On the basis of these assumptions, we conclude (see Section A.2) that the physical components $(\nabla \boldsymbol{\Psi})_{\alpha\beta\gamma} \equiv (\nabla \boldsymbol{\Psi})_{\alpha(\beta\gamma)}$ have the explicit form as follows:

$$(\nabla \boldsymbol{\Psi})_{rrr} = \partial_r \Psi_{rr}, \quad (\nabla \boldsymbol{\Psi})_{r\varphi\varphi} = \partial_r \Psi_{\varphi\varphi}, \quad (\nabla \boldsymbol{\Psi})_{rr\varphi} = \partial_r \Psi_{r\varphi}, \tag{39}$$

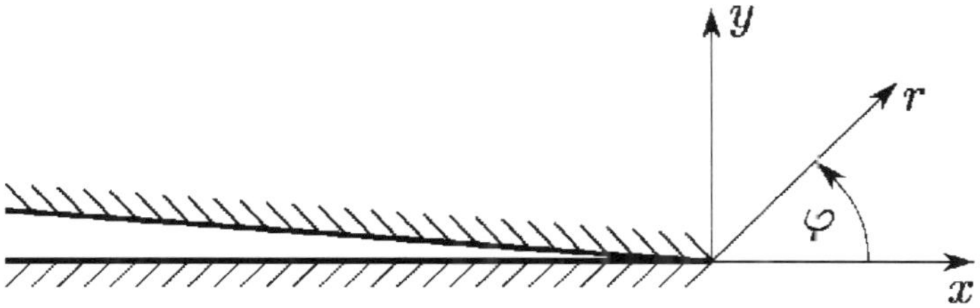

Figure 1.
Coordinate axes ahead of the crack tip.

$$(\nabla\boldsymbol{\Psi})_{\varphi rr} = \frac{1}{r}\left(\partial_\varphi \Psi_{rr} - 2\Psi_{r\varphi}\right), \tag{40}$$

$$(\nabla\boldsymbol{\Psi})_{\varphi\varphi\varphi} = \frac{1}{r}\left(\partial_\varphi \Psi_{\varphi\varphi} + 2\Psi_{r\varphi}\right), \tag{41}$$

$$(\nabla\boldsymbol{\Psi})_{\varphi r\varphi} = \frac{1}{r}\left(\partial_\varphi \Psi_{r\varphi} + \Psi_{rr} - \Psi_{\varphi\varphi}\right), \tag{42}$$

whereas all other components of $\nabla\boldsymbol{\Psi}$ vanish,

$$(\nabla\boldsymbol{\Psi})_{\alpha\beta z} = (\nabla\boldsymbol{\Psi})_{z\alpha\beta} = 0. \tag{43}$$

Similarly, we find (see Section 6.3) for the physical components $(\Delta\boldsymbol{\Psi})_{\alpha\beta} = (\Delta\boldsymbol{\Psi})_{(\alpha\beta)}$, that

$$(\Delta\boldsymbol{\Psi})_{rr} = \partial_{rr}\Psi_{rr} + \frac{1}{r^2}\partial_{\varphi\varphi}\Psi_{rr} + \frac{1}{r}\partial_r\Psi_{rr} - \frac{4}{r^2}\partial_\varphi\Psi_{r\varphi} - \frac{2}{r^2}\Psi_{rr} + \frac{2}{r^2}\Psi_{\varphi\varphi}, \tag{44}$$

$$(\Delta\boldsymbol{\Psi})_{\varphi\varphi} = \partial_{rr}\Psi_{\varphi\varphi} + \frac{1}{r^2}\partial_{\varphi\varphi}\Psi_{\varphi\varphi} + \frac{1}{r}\partial_r\Psi_{\varphi\varphi} + \frac{4}{r^2}\partial_\varphi\Psi_{r\varphi} + \frac{2}{r^2}\Psi_{rr} - \frac{2}{r^2}\Psi_{\varphi\varphi}, \tag{45}$$

$$(\Delta\boldsymbol{\Psi})_{r\varphi} = \partial_{rr}\Psi_{r\varphi} + \frac{1}{r^2}\partial_{\varphi\varphi}\Psi_{r\varphi} + \frac{1}{r}\partial_r\Psi_{r\varphi} + \frac{2}{r^2}\partial_\varphi\Psi_{rr} - \frac{2}{r^2}\partial_\varphi\Psi_{\varphi\varphi} - \frac{4}{r^2}\Psi_{r\varphi}, \tag{46}$$

$$(\Delta\boldsymbol{\Psi})_{\alpha z} = 0. \tag{47}$$

It is well known (see, e.g., Anderson [12], p. 114) that the nonvanishing physical components of ε are given by the following expressions:

$$\varepsilon_{rr} = \partial_r u_r, \quad \varepsilon_{\varphi\varphi} = \frac{1}{r}\left(u_r + \partial_\varphi u_\varphi\right), \quad \varepsilon_{r\varphi} = \frac{1}{2}\left(\frac{1}{r}\partial_\varphi u_r + \partial_r u_\varphi - \frac{1}{r}u_\varphi\right). \tag{48}$$

4.2 Cauchy stress: classical equilibrium equations

In view of the assumptions of the last section, we may derive the following results. We conclude from Eqs. (21)–(23), that

$$\Sigma_{\alpha\beta} = \Sigma_{\alpha\beta}(r,\varphi), \quad \tau_{\alpha\beta} = \tau_{\alpha\beta}(r,\varphi), \quad \sigma_{\alpha\beta} = \sigma_{\alpha\beta}(r,\varphi), \tag{49}$$

and that

$$[\Sigma_{\alpha\beta}] = \begin{pmatrix} \Sigma_{rr} & \Sigma_{r\varphi} & 0 \\ \Sigma_{r\varphi} & \Sigma_{\varphi\varphi} & 0 \\ 0 & 0 & \Sigma_{zz} \end{pmatrix}, \quad [\tau_{\alpha\beta}] = \begin{pmatrix} \tau_{rr} & \tau_{r\varphi} & 0 \\ \tau_{r\varphi} & \tau_{\varphi\varphi} & 0 \\ 0 & 0 & \tau_{zz} \end{pmatrix}, \tag{50}$$

$$[\sigma_{\alpha\beta}] = \begin{pmatrix} \sigma_{rr} & \sigma_{r\varphi} & 0 \\ \sigma_{r\varphi} & \sigma_{\varphi\varphi} & 0 \\ 0 & 0 & \sigma_{zz} \end{pmatrix}. \tag{51}$$

Since $\varepsilon_{zz} = \Psi_{zz} = 0$, we deduce from the elasticity laws (21)–(23) that

$$\Sigma_{zz} = \nu\left(\Sigma_{rr} + \Sigma_{\varphi\varphi}\right), \quad \tau_{zz} = \nu\left(\tau_{rr} + \tau_{\varphi\varphi}\right), \quad \sigma_{zz} = \nu\left(\sigma_{rr} + \sigma_{\varphi\varphi}\right), \tag{52}$$

where ν is given by Eq. (9). Thus, as in classical elasticity, once Σ_{rr} and $\Sigma_{\varphi\varphi}$ have been determined, Σ_{zz} is calculated by the first equation of (52).

Quite similar to the case of classical elasticity (see, e.g., Anderson [12], p. 114), the matrix of the components of $\boldsymbol{\Sigma}$ in Eq. (50) and the classical equilibrium condition (24), expressed in physical components, lead to the following two equations:

$$\partial_r \Sigma_{rr} + \frac{1}{r}\partial_\varphi \Sigma_{r\varphi} + \frac{1}{r}\left(\Sigma_{rr} - \Sigma_{\varphi\varphi}\right) = 0, \tag{53}$$

$$\partial_r \Sigma_{r\varphi} + \frac{1}{r}\partial_\varphi \Sigma_{\varphi\varphi} + \frac{2}{r}\Sigma_{r\varphi} = 0. \tag{54}$$

4.3 Classical compatibility condition

For the analytical solutions in Part *II*, we need the classical compatibility condition for the components of the strain tensor ε (see, e.g., Malvern [13], p. 669)

$$\partial_{rr}\varepsilon_{\varphi\varphi} + \frac{1}{r^2}\partial_{\varphi\varphi}\varepsilon_{rr} - \frac{2}{r}\partial_{r\varphi}\varepsilon_{r\varphi} - \frac{1}{r}\partial_r\varepsilon_{rr} + \frac{2}{r}\partial_r\varepsilon_{\varphi\varphi} - \frac{2}{r^2}\partial_\varphi\varepsilon_{r\varphi} = 0. \tag{55}$$

The aim is now to rewrite this equation in terms of the physical components of $\boldsymbol{\Sigma}$ and $\boldsymbol{\Psi}$ by using Eq. (34). There are various equivalent representations for $\mathbb{C}^{-1}$, depending on the chosen set of elasticity constants. We find it convenient here to express $\mathbb{C}^{-1}$ in terms of the elasticity constants μ and ν. Thus, from Eq. (34), expressed in physical components,

$$\varepsilon_{rr} = \frac{c_1}{2\mu c_2}\left[\Sigma_{rr} - \nu(\Sigma_{rr} + \Sigma_{\varphi\varphi})\right] + \frac{c_2 - c_1}{c_1}\Psi_{rr}, \tag{56}$$

$$\varepsilon_{\varphi\varphi} = \frac{c_1}{2\mu c_2}\left[\Sigma_{\varphi\varphi} - \nu(\Sigma_{rr} + \Sigma_{\varphi\varphi})\right] + \frac{c_2 - c_1}{c_1}\Psi_{\varphi\varphi}, \tag{57}$$

$$\varepsilon_{r\varphi} = \frac{c_1}{2\mu c_2}\Sigma_{r\varphi} + \frac{c_2 - c_1}{c_1}\Psi_{r\varphi}. \tag{58}$$

By inserting these equations into Eq. (55), we get

$$\begin{aligned}
&\partial_{rr}\left(\frac{c_1}{2\mu c_2}\left[\Sigma_{\varphi\varphi} - \nu\left(\Sigma_{rr} + \Sigma_{\varphi\varphi}\right)\right] + \frac{c_2 - c_1}{c_2}\Psi_{\varphi\varphi}\right)\\
&-\frac{2}{r}\partial_{r\varphi}\left(\frac{c_1}{2\mu c_2}\Sigma_{r\varphi} + \frac{c_2 - c_1}{c_2}\Psi_{r\varphi}\right)\\
&+\frac{1}{r^2}\partial_{\varphi\varphi}\left(\frac{c_1}{2\mu c_2}\left[\Sigma_{rr} - \nu\left(\Sigma_{rr} + \Sigma_{\varphi\varphi}\right)\right] + \frac{c_2 - c_1}{c_2}\Psi_{rr}\right)\\
&-\frac{1}{r}\partial_r\left(\frac{c_1}{2\mu c_2}\left[\Sigma_{rr} - \nu\left(\Sigma_{rr} + \Sigma_{\varphi\varphi}\right)\right] + \frac{c_2 - c_1}{c_2}\Psi_{rr}\right)\\
&+\frac{2}{r}\partial_r\left(\frac{c_1}{2\mu c_2}\left[\Sigma_{\varphi\varphi} - \nu\left(\Sigma_{rr} + \Sigma_{\varphi\varphi}\right)\right] + \frac{c_2 - c_1}{c_2}\Psi_{\varphi\varphi}\right)\\
&-\frac{2}{r^2}\partial_\varphi\left(\frac{c_1}{2\mu c_2}\Sigma_{r\varphi} + \frac{c_2 - c_1}{c_2}\Psi_{r\varphi}\right) = 0.
\end{aligned} \tag{59}$$

This is equivalent to a vanishing sum of two functions of $\Psi_{\alpha\beta}$ and $\Sigma_{\alpha\beta}$, respectively:

$$0 = \chi_1(\Psi_{\alpha\beta}) + \chi_2(\Sigma_{\alpha\beta}), \tag{60}$$

with

$$\chi_1(\Psi_{\alpha\beta}) := \frac{c_2 - c_1}{c_2}\left[\partial_{rr}\Psi_{\varphi\varphi} - \frac{2}{r}\partial_{r\varphi}\Psi_{r\varphi} + \frac{1}{r^2}\partial_{\varphi\varphi}\Psi_{rr}\right. \\ \left. - \frac{1}{r}\partial_r\Psi_{rr} + \frac{2}{r}\partial_r\Psi_{\varphi\varphi} - \frac{2}{r^2}\partial_\varphi\Psi_{r\varphi}\right] \tag{61}$$

and

$$\chi_2(\Sigma_{\alpha\beta}) := \frac{c_1}{2\mu c_2}\left[\partial_{rr}\Sigma_{\varphi\varphi} - \nu\,\partial_{rr}(\Sigma_{rr} + \Sigma_{\varphi\varphi}) - \frac{2}{r}\partial_{r\varphi}\Sigma_{r\varphi} + \frac{1}{r^2}\partial_{\varphi\varphi}\Sigma_{rr}\right. \\ - \frac{\nu}{r^2}\partial_{\varphi\varphi}(\Sigma_{rr} + \Sigma_{\varphi\varphi}) - \frac{1}{r}\partial_r\Sigma_{rr} \\ \left. - \frac{\nu}{r}\partial_r(\Sigma_{rr} + \Sigma_{\varphi\varphi}) + \frac{2}{r}\partial_r\Sigma_{\varphi\varphi} - \frac{2}{r^2}\partial_\varphi\Sigma_{r\varphi}\right], \tag{62}$$

The right hand side of Eq. (62) can be simplified by invoking the equilibrium Eqs. (53) and (54). First recast Eq. (54) to solve for $\Sigma_{r\varphi}$,

$$\Sigma_{r\varphi} = -\frac{r}{2}\partial_r\Sigma_{r\varphi} - \frac{1}{2}\partial_\varphi\Sigma_{\varphi\varphi}, \tag{63}$$

and then take the derivative with respect to φ,

$$\partial_\varphi\Sigma_{r\varphi} = -\frac{r}{2}\partial_{r\varphi}\Sigma_{r\varphi} - \frac{1}{2}\partial_{\varphi\varphi}\Sigma_{\varphi\varphi}. \tag{64}$$

On the other hand, from Eq. (53),

$$\partial_\varphi\Sigma_{r\varphi} = -r\,\partial_r\Sigma_{rr} - \Sigma_{rr} + \Sigma_{\varphi\varphi}, \tag{65}$$

and, after differentiation with respect to r,

$$\partial_{r\varphi}\Sigma_{r\varphi} = -r\,\partial_{rr}\Sigma_{rr} - 2\partial_r\Sigma_{rr} + \partial_r\Sigma_{\varphi\varphi}. \tag{66}$$

By substituting the latter into Eq. (64),

$$\partial_\varphi\Sigma_{r\varphi} = \frac{r^2}{2}\partial_{rr}\Sigma_{rr} - \frac{1}{2}\partial_{\varphi\varphi}\Sigma_{\varphi\varphi} + r\,\partial_r\Sigma_{rr} - \frac{r}{2}\partial_r\Sigma_{\varphi\varphi}. \tag{67}$$

Finally, by substituting Eqs. (66) and (67) into Eq. (62) and after some rearrangement of terms, we find that

$$\chi_2(\Sigma_{\alpha\beta}) = \frac{(1-\nu)c_1}{2\mu c_2}\left[\partial_{rr}(\Sigma_{rr} + \Sigma_{\varphi\varphi}) + \frac{1}{r^2}\partial_{\varphi\varphi}(\Sigma_{rr} + \Sigma_{\varphi\varphi})\right. \\ \left. + \frac{1}{r}\partial_r(\Sigma_{rr} + \Sigma_{\varphi\varphi})\right]. \tag{68}$$

4.4 Field equations for Ψ

By expressing $\mathbb{C}^{-1}$ in terms of μ and ν and using Eqs. (50), (37) and (44)–(46), we can readily obtain from Eq. (35), expressed in physical components, the following three equations:

$$\partial_{rr}\Psi_{rr} + \frac{1}{r^2}\partial_{\varphi\varphi}\Psi_{rr} + \frac{1}{r}\partial_r\Psi_{rr} - \frac{4}{r^2}\partial_\varphi\Psi_{r\varphi} - \left(\frac{2}{r^2} + \frac{1}{c_2}\right)\Psi_{rr} + \frac{2}{r^2}\Psi_{\varphi\varphi} + \frac{1-\nu}{2\mu c_2}\Sigma_{rr} - \frac{\nu}{2\mu c_2}\Sigma_{\varphi\varphi} = 0, \tag{69}$$

$$\partial_{rr}\Psi_{\varphi\varphi} + \frac{1}{r^2}\partial_{\varphi\varphi}\Psi_{\varphi\varphi} + \frac{1}{r}\partial_r\Psi_{\varphi\varphi} + \frac{4}{r^2}\partial_\varphi\Psi_{r\varphi} + \frac{2}{r^2}\Psi_{rr} - \left(\frac{2}{r^2} + \frac{1}{c_2}\right)\Psi_{\varphi\varphi} + \frac{1-\nu}{2\mu c_2}\Sigma_{\varphi\varphi} - \frac{\nu}{2\mu c_2}\Sigma_{rr} = 0, \tag{70}$$

$$\partial_{rr}\Psi_{r\varphi} + \frac{1}{r^2}\partial_{\varphi\varphi}\Psi_{r\varphi} + \frac{1}{r}\partial_r\Psi_{r\varphi} + \frac{2}{r^2}\partial_\varphi\Psi_{rr} - \frac{2}{r^2}\partial_\varphi\Psi_{\varphi\varphi} - \left(\frac{4}{r^2} + \frac{1}{c_2}\right)\Psi_{r\varphi} + \frac{1}{2\mu c_2}\Sigma_{r\varphi} = 0. \tag{71}$$

4.5 Double stress

4.5.1 Elasticity law for double stress

With respect to physical components, the elasticity law (24) becomes

$$\begin{aligned}\mu_{\alpha\beta\gamma} \equiv \mu_{\alpha(\beta\gamma)} &= (c_2 - c_1)(\nabla\boldsymbol{\Psi})_{\alpha\rho\zeta}\mathbb{C}_{\rho\zeta\beta\gamma} \\ &= (c_2 - c_1)\Big\{2\mu(\nabla\boldsymbol{\Psi})_{\alpha\beta\gamma} + \lambda(\nabla\boldsymbol{\Psi})_{\alpha\zeta\zeta}\delta_{\beta\gamma}\Big\}.\end{aligned} \tag{72}$$

Keeping in mind that the physical components of $\nabla\boldsymbol{\Psi}$ for a plane strain state are given by Eqs. (39)–(41), it is not difficult to derive the following results:

$$\mu_{rrr} = (c_2 - c_1)\left[(\lambda + 2\mu)\partial_r\Psi_{rr} + \lambda\partial_r\Psi_{\varphi\varphi}\right], \tag{73}$$

$$\mu_{r\varphi\varphi} = (c_2 - c_1)\left[(\lambda + 2\mu)\partial_r\Psi_{\varphi\varphi} + \lambda\partial_r\Psi_{rr}\right], \tag{74}$$

$$\mu_{rzz} = (c_2 - c_1)\lambda\partial_r(\Psi_{rr} + \Psi_{\varphi\varphi}) = \nu(\mu_{rrr} + \mu_{r\varphi\varphi}), \tag{75}$$

$$\mu_{rr\varphi} = (c_2 - c_1)2\mu\partial_r\Psi_{r\varphi}, \tag{76}$$

$$\mu_{\varphi rr} = \frac{c_2 - c_1}{r}\left[2\mu(\partial_\varphi\Psi_{rr} - 2\Psi_{r\varphi}) + \lambda\partial_\varphi(\Psi_{rr} + \Psi_{\varphi\varphi})\right], \tag{77}$$

$$\mu_{\varphi\varphi\varphi} = \frac{c_2 - c_1}{r}\left[2\mu(\partial_\varphi\Psi_{\varphi\varphi} + 2\Psi_{r\varphi}) + \lambda\partial_\varphi(\Psi_{rr} + \Psi_{\varphi\varphi})\right], \tag{78}$$

$$\mu_{\varphi zz} = \frac{c_2 - c_1}{r}\lambda\partial_\varphi(\Psi_{rr} + \Psi_{\varphi\varphi}) = \nu(\mu_{\varphi rr} + \mu_{\varphi\varphi\varphi}), \tag{79}$$

$$\mu_{\varphi r\varphi} = \frac{c_2 - c_1}{r}2\mu(\partial_\varphi\Psi_{r\varphi} + \Psi_{rr} - \Psi_{\varphi\varphi}), \tag{80}$$

$$\mu_{rrz} = \mu_{r\varphi z} = \mu_{\varphi rz} = \mu_{\varphi\varphi z} = 0, \tag{81}$$

$$\mu_{z\alpha\beta} = 0. \tag{82}$$

4.5.2 Non-classical equilibrium conditions

The physical components of the non-classical equilibrium condition (26) are

$$(\operatorname{div} \boldsymbol{\mu})_{\alpha\beta} + \sigma_{\alpha\beta} = 0. \tag{83}$$

With the aid of the physical components of $\boldsymbol{\mu}$ given by Eqs. (73)–(81) and the physical components of σ stated by Eq. (51), we can verify that the equilibrium conditions (83) furnish only four nontrivial equations:

$$(\operatorname{div} \boldsymbol{\mu})_{rr} + \sigma_{rr} = 0, \tag{84}$$

$$(\operatorname{div} \boldsymbol{\mu})_{\varphi\varphi} + \sigma_{\varphi\varphi} = 0, \tag{85}$$

$$(\operatorname{div} \boldsymbol{\mu})_{r\varphi} + \sigma_{r\varphi} = 0, \tag{86}$$

$$(\operatorname{div} \boldsymbol{\mu})_{zz} + \sigma_{zz} = 0, \tag{87}$$

or equivalently (cf. Section A.4)

$$\partial_r \mu_{rrr} + \frac{1}{r} \partial_\varphi \mu_{\varphi rr} + \frac{1}{r} \left(\mu_{rrr} - 2\mu_{\varphi r\varphi}\right) + \sigma_{rr} = 0, \tag{88}$$

$$\partial_r \mu_{r\varphi\varphi} + \frac{1}{r} \partial_\varphi \mu_{\varphi\varphi\varphi} + \frac{1}{r} \left(\mu_{r\varphi\varphi} + 2\mu_{\varphi r\varphi}\right) + \sigma_{\varphi\varphi} = 0, \tag{89}$$

$$\partial_r \mu_{rr\varphi} + \frac{1}{r} \partial_\varphi \mu_{\varphi r\varphi} + \frac{1}{r} \left(\mu_{rr\varphi} - \mu_{\varphi\varphi\varphi} + \mu_{\varphi rr}\right) + \sigma_{r\varphi} = 0, \tag{90}$$

$$\partial_r \mu_{rzz} + \frac{1}{r} \partial_\varphi \mu_{\varphi zz} + \frac{1}{r} \mu_{rzz} + \sigma_{zz} = 0. \tag{91}$$

4.6 Nonclassical compatibility conditions

Besides the classical compatibility condition for the strain ε in Eq. (55), further compatibility conditions for the micro-strain $\boldsymbol{\Psi}$ can be established by considering the following identities:

$$\partial_{r\varphi} \Psi_{rr} - \partial_{\varphi r} \Psi_{rr} = 0, \tag{92}$$

$$\partial_{r\varphi} \Psi_{\varphi\varphi} - \partial_{\varphi r} \Psi_{\varphi\varphi} = 0, \tag{93}$$

$$\partial_{r\varphi} \Psi_{r\varphi} - \partial_{\varphi r} \Psi_{r\varphi} = 0. \tag{94}$$

From these, we obtain useful relations by involving the physical components $\mu_{\alpha\beta\gamma}$ with the aid of the elasticity law (24). To illustrate, we recall from Eqs. (39)–(43) that

$$\begin{aligned} \partial_{r\varphi} \Psi_{rr} &= \partial_r \left(\partial_\varphi \Psi_{rr}\right) = \partial_r \left(r (\nabla \boldsymbol{\Psi})_{\varphi rr} + 2\Psi_{r\varphi} \right) \\ &= (\nabla \boldsymbol{\Psi})_{\varphi rr} + r \partial_r (\nabla \boldsymbol{\Psi})_{\varphi rr} + 2(\nabla \boldsymbol{\Psi})_{rr\varphi} \end{aligned} \tag{95}$$

and that

$$\partial_{\varphi r} \Psi_{rr} = \partial_\varphi (\partial_r \Psi_{rr}) = \partial_\varphi (\nabla \boldsymbol{\Psi})_{rrr}. \tag{96}$$

By inserting these into Eq. (92), we obtain the following equation:

$$\partial_\varphi (\nabla \boldsymbol{\Psi})_{rrr} - (\nabla \boldsymbol{\Psi})_{\varphi rr} - r \partial_r (\nabla \boldsymbol{\Psi})_{\varphi rr} - 2(\nabla \boldsymbol{\Psi})_{rr\varphi} = 0. \tag{97}$$

In a similar way, we conclude from Eqs. (93) and (94) that

$$\partial_\varphi(\nabla\boldsymbol{\Psi})_{r\varphi\varphi} - (\nabla\boldsymbol{\Psi})_{\varphi\varphi\varphi} - r\,\partial_r(\nabla\boldsymbol{\Psi})_{\varphi\varphi\varphi} + 2(\nabla\boldsymbol{\Psi})_{rr\varphi} = 0, \tag{98}$$

$$\partial_\varphi(\nabla\boldsymbol{\Psi})_{rr\varphi} - (\nabla\boldsymbol{\Psi})_{\varphi r\varphi} - r\,\partial_r(\nabla\boldsymbol{\Psi})_{\varphi r\varphi} + (\nabla\boldsymbol{\Psi})_{rrr} - (\nabla\boldsymbol{\Psi})_{r\varphi\varphi} = 0. \tag{99}$$

In order to involve the components of $\boldsymbol{\mu}$, we invert Eq. (24) to obtain

$$(\nabla\boldsymbol{\Psi})_{\alpha\beta\gamma} = \frac{1}{(c_2 - c_1)E}\left((1+\nu)\,\mu_{\alpha\beta\gamma} - \nu\,\mu_{\alpha\zeta\zeta}\,\delta_{\beta\gamma}\right), \tag{100}$$

where E is given by Eq. (9). Explicitly, we get

$$(\nabla\boldsymbol{\Psi})_{rrr} = \frac{1}{(c_2 - c_1)E}\left((1-\nu^2)\,\mu_{rrr} - \nu(1+\nu)\,\mu_{r\varphi\varphi}\right), \tag{101}$$

$$(\nabla\boldsymbol{\Psi})_{r\varphi\varphi} = \frac{1}{(c_2 - c_1)E}\left((1-\nu^2)\,\mu_{r\varphi\varphi} - \nu(1+\nu)\,\mu_{rrr}\right), \tag{102}$$

$$(\nabla\boldsymbol{\Psi})_{rr\varphi} = \frac{1+\nu}{(c_2 - c_1)E}\,\mu_{rr\varphi}, \tag{103}$$

$$(\nabla\boldsymbol{\Psi})_{\varphi rr} = \frac{1}{(c_2 - c_1)E}\left((1-\nu^2)\,\mu_{\varphi rr} - \nu(1+\nu)\,\mu_{\varphi\varphi\varphi}\right), \tag{104}$$

$$(\nabla\boldsymbol{\Psi})_{\varphi\varphi\varphi} = \frac{1}{(c_2 - c_1)E}\left((1-\nu^2)\,\mu_{\varphi\varphi\varphi} - \nu(1+\nu)\,\mu_{\varphi rr}\right), \tag{105}$$

$$(\nabla\boldsymbol{\Psi})_{\varphi r\varphi} = \frac{1+\nu}{(c_2 - c_1)E}\,\mu_{\varphi r\varphi}, \tag{106}$$

where in addition use has been made of Eqs. (75) and (80). By inserting these components into Eqs. (97)–(99), we can verify that

$$\begin{aligned} &-(1-\nu^2)\,r\,\partial_r\mu_{\varphi rr} + \nu(1+\nu)\,r\,\partial_r\mu_{\varphi\varphi\varphi} + (1-\nu^2)\,\partial_\varphi\mu_{rrr} - \nu(1+\nu)\,\partial_\varphi\mu_{r\varphi\varphi} \\ &-(1-\nu^2)\,\mu_{\varphi rr} + \nu(1+\nu)\,\mu_{\varphi\varphi\varphi} - 2(1+\nu)\,\mu_{rr\varphi} = 0, \end{aligned} \tag{107}$$

$$\begin{aligned} &-(1-\nu^2)\,r\,\partial_r\mu_{\varphi\varphi\varphi} + \nu(1+\nu)\,r\,\partial_r\mu_{\varphi rr} + (1-\nu^2)\,\partial_\varphi\mu_{r\varphi\varphi} - \nu(1+\nu)\,\partial_\varphi\mu_{rrr} \\ &-(1-\nu^2)\,\mu_{\varphi\varphi\varphi} + \nu(1+\nu)\,\mu_{\varphi rr} + 2(1+\nu)\,\mu_{rr\varphi} = 0, \end{aligned} \tag{108}$$

$$-r\,\partial_r\mu_{\varphi r\varphi} + \partial_\varphi\mu_{rr\varphi} - \mu_{\varphi r\varphi} + \mu_{rrr} - \mu_{r\varphi\varphi} = 0. \tag{109}$$

The last equation is independent of material parameters. In order to rewrite Eqs. (107) and (108) also in a form independent of material parameters, we add and subtract them from each other to obtain, respectively,

$$\partial_\varphi\mu_{r\varphi\varphi} + \partial_\varphi\mu_{rrr} - \mu_{\varphi\varphi\varphi} - \mu_{\varphi rr} - r\,\partial_r\mu_{\varphi\varphi\varphi} - r\,\partial_r\mu_{\varphi rr} = 0, \tag{110}$$

$$\partial_\varphi\mu_{r\varphi\varphi} - \partial_\varphi\mu_{rrr} - \mu_{\varphi\varphi\varphi} + \mu_{\varphi rr} - r\,\partial_r\mu_{\varphi\varphi\varphi} + r\,\partial_r\mu_{\varphi rr} + 4\mu_{rr\varphi} = 0. \tag{111}$$

4.7 Boundary conditions

As usually, near-tip field solutions rely upon boundary conditions, which are imposed only on the crack faces. Especially, we assume the classical traction $\boldsymbol{P}$ (see Eq. (27)) and the double force $\boldsymbol{T}$ (see Eq. (28)) to vanish on the crack faces. With regard to **Figure 1** this implies that

$$[P_\alpha]_{\varphi=\pm\pi} = 0, \quad [T_{\alpha\beta}]_{\varphi=\pm\pi} = 0. \tag{112}$$

Now, we have from Eq. (27), expressed in physical components, that $P_\alpha = n_\beta \Sigma_{\beta\alpha}$, where on the crack faces $[\boldsymbol{n}]_{\varphi=\pm\pi} = \mp \boldsymbol{e}_\varphi$. Therefore, and by virtue of Eq. (50), the nontrivial classical boundary conditions implied by Eq. (112) are as follows:

$$[\Sigma_{r\varphi}]_{\varphi=\pm\pi} = 0, \tag{113}$$

$$[\Sigma_{\varphi\varphi}]_{\varphi=\pm\pi} = 0. \tag{114}$$

Similarly, we get from Eq. (28), expressed in physical components, that $T_{\alpha\beta} = n_\gamma \mu_{\gamma\alpha\beta}$, so that the nonclassical part of Eq. (112) implies

$$[\mu_{\varphi\alpha\beta}]_{\varphi=\pm\pi} = (c_2 - c_1)\left[(\nabla\boldsymbol{\Psi})_{\varphi\gamma\zeta}\right]_{\varphi=\pm\pi} \mathbb{C}_{\gamma\zeta\alpha\beta} = 0, \tag{115}$$

where use has been made of the elasticity law (24). As the isotropic elasticity tensor $\mathbb{C}$ has been assumed to be invertible, we infer from Eq. (115), that

$$0 = \left[(\nabla\boldsymbol{\Psi})_{\varphi\alpha\beta}\right]_{\varphi=\pm\pi}. \tag{116}$$

Keeping in mind Eqs. (39)–(43), the only nontrivial conditions implied are as follows:

$$\left[(\nabla\boldsymbol{\Psi})_{\varphi rr}\right]_{\varphi=\pm\pi} = \left[(\nabla\boldsymbol{\Psi})_{\varphi\varphi\varphi}\right]_{\varphi=\pm\pi} = \left[(\nabla\boldsymbol{\Psi})_{\varphi r\varphi}\right]_{\varphi=\pm\pi} = 0, \tag{117}$$

or equivalently

$$[\partial_\varphi \Psi_{rr} - 2\Psi_{r\varphi}]_{\varphi=\pm\pi} = 0, \tag{118}$$

$$[\partial_\varphi \Psi_{\varphi\varphi} + 2\Psi_{r\varphi}]_{\varphi=\pm\pi} = 0, \tag{119}$$

$$[\partial_\varphi \Psi_{r\varphi} + \Psi_{rr} - \Psi_{\varphi\varphi}]_{\varphi=\pm\pi} = 0. \tag{120}$$

4.8 Symmetry conditions

Symmetry conditions are important to classify the near-tip field solutions into types according to Mode-I and Mode-II crack problems. Each type of loading condition is characterized by the following symmetry conditions.

4.8.1 Mode-I

As in classical elasticity (see, e.g., Hellan [14], p. 10), we suppose for the macro-displacement the following symmetry conditions:

$$u_r(r,\varphi) = u_r(r,-\varphi), \quad u_\varphi(r,\varphi) = -u_\varphi(r,-\varphi), \tag{121}$$

i.e., u_r is an even function of φ, whereas u_φ is an odd function of φ. It follows from Eq. (48) that the physical components of the macro-strain ε exhibit the following properties:

$$\varepsilon_{rr}(r,\varphi) = \varepsilon_{rr}(r,-\varphi), \quad \varepsilon_{\varphi\varphi}(r,\varphi) = \varepsilon_{\varphi\varphi}(r,-\varphi), \tag{122}$$

$$\varepsilon_{r\varphi}(r,\varphi) = -\varepsilon_{r\varphi}(r,-\varphi), \tag{123}$$

implying that ε_{rr} and $\varepsilon_{\varphi\varphi}$ are even functions of φ, whereas $\varepsilon_{r\varphi}$ is an odd function of φ. Since $\boldsymbol{\Psi}$ is also a strain tensor, we assume for its components, in analogy to Eqs. (122) and (123), that

$$\Psi_{rr}(r,\varphi) = \Psi_{rr}(r,-\varphi), \quad \Psi_{\varphi\varphi}(r,\varphi) = \Psi_{\varphi\varphi}(r,-\varphi), \tag{124}$$

$$\Psi_{r\varphi}(r,\varphi) = -\Psi_{r\varphi}(r,-\varphi). \tag{125}$$

Then, it can be verified, with the help of the elasticity law (21), expressed in physical components, that Eqs. (122)–(125) engender the following conditions for the components of $\boldsymbol{\Sigma}$:

$$\Sigma_{rr}(r,\varphi) = \Sigma_{rr}(r,-\varphi), \quad \Sigma_{\varphi\varphi}(r,\varphi) = \Sigma_{\varphi\varphi}(r,-\varphi), \tag{126}$$

$$\Sigma_{r\varphi}(r,\varphi) = -\Sigma_{r\varphi}(r,-\varphi). \tag{127}$$

Further, it can be seen from Eqs. (124) and (125), that

$$\partial_r\Psi_{rr}(r,\varphi) = \partial_r\Psi_{rr}(r,-\varphi), \; \partial_\varphi\Psi_{rr}(r,\varphi) = -\partial_\varphi\Psi_{rr}(r,-\varphi), \tag{128}$$

$$\partial_r\Psi_{\varphi\varphi}(r,\varphi) = \partial_r\Psi_{\varphi\varphi}(r,-\varphi), \; \partial_\varphi\Psi_{\varphi\varphi}(r,\varphi) = -\partial_\varphi\Psi_{\varphi\varphi}(r,-\varphi), \tag{129}$$

$$\partial_r\Psi_{r\varphi}(r,\varphi) = \partial_r\Psi_{r\varphi}(r,-\varphi), \; \partial_\varphi\Psi_{r\varphi}(r,\varphi) = -\partial_\varphi\Psi_{r\varphi}(r,-\varphi), \tag{130}$$

and from the elasticity laws (73)–(82), that

$$\mu_{rrr}(r,\varphi) = \mu_{rrr}(r,-\varphi), \quad \mu_{\varphi rr}(r,\varphi) = -\mu_{\varphi rr}(r,-\varphi), \tag{131}$$

$$\mu_{r\varphi\varphi}(r,\varphi) = \mu_{r\varphi\varphi}(r,-\varphi), \quad \mu_{\varphi\varphi\varphi}(r,\varphi) = -\mu_{\varphi\varphi\varphi}(r,-\varphi), \tag{132}$$

$$\mu_{rzz}(r,\varphi) = \mu_{rzz}(r,-\varphi), \quad \mu_{\varphi zz}(r,\varphi) = -\mu_{\varphi zz}(r,-\varphi), \tag{133}$$

$$\mu_{rr\varphi}(r,\varphi) = \mu_{rr\varphi}(r,-\varphi), \quad \mu_{\varphi r\varphi}(r,\varphi) = -\mu_{\varphi r\varphi}(r,-\varphi). \tag{134}$$

4.8.2 Mode-II

We know from classical elasticity (see, e.g., Hellan [14], p. 10), that the radial component of the displacement vector is an odd function of φ, whereas the circumferential component is an even function of φ. We assume these symmetry properties to also apply for the macro-displacement here, i.e.,

$$u_r(r,\varphi) = -u_r(r,-\varphi), \quad u_\varphi(r,\varphi) = u_\varphi(r,-\varphi). \tag{135}$$

It follows for the macro-strain ε, that

$$\varepsilon_{rr}(r,\varphi) = -\varepsilon_{rr}(r,-\varphi), \quad \varepsilon_{\varphi\varphi}(r,\varphi) = -\varepsilon_{\varphi\varphi}(r,-\varphi), \tag{136}$$

$$\varepsilon_{r\varphi}(r,\varphi) = \varepsilon_{r\varphi}(r,-\varphi), \tag{137}$$

which suggest to assume the following symmetries for $\boldsymbol{\Psi}$:

$$\Psi_{rr}(r,\varphi) = -\Psi_{rr}(r,-\varphi), \quad \Psi_{\varphi\varphi}(r,\varphi) = -\Psi_{\varphi\varphi}(r,-\varphi), \tag{138}$$

$$\Psi_{r\varphi}(r,\varphi) = \Psi_{r\varphi}(r,-\varphi). \tag{139}$$

It can be proved, in a similar fashion to Mode-I, that

$$\Sigma_{rr}(r,\varphi) = -\Sigma_{rr}(r,-\varphi), \quad \Sigma_{\varphi\varphi}(r,\varphi) = -\Sigma_{\varphi\varphi}(r,-\varphi), \tag{140}$$

$$\Sigma_{r\varphi}(r,\varphi) = \Sigma_{r\varphi}(r,-\varphi), \tag{141}$$

and that

$$\mu_{rrr}(r,\varphi) = -\mu_{rrr}(r,-\varphi), \quad \mu_{\varphi rr}(r,\varphi) = \mu_{\varphi rr}(r,-\varphi), \tag{142}$$

$$\mu_{r\varphi\varphi}(r,\varphi) = -\mu_{r\varphi\varphi}(r,-\varphi), \quad \mu_{\varphi\varphi\varphi}(r,\varphi) = \mu_{\varphi\varphi\varphi}(r,-\varphi), \tag{143}$$

$$\mu_{rzz}(r,\varphi) = -\mu_{rzz}(r,-\varphi), \quad \mu_{\varphi zz}(r,\varphi) = \mu_{\varphi zz}(r,-\varphi), \tag{144}$$

$$\mu_{rr\varphi}(r,\varphi) = -\mu_{rr\varphi}(r,-\varphi), \quad \mu_{\varphi r\varphi}(r,\varphi) = \mu_{\varphi r\varphi}(r,-\varphi). \tag{145}$$

Before closing this section, we notice here, that the numerical simulations on the basis of the finite element method in Part *III* confirm the assumed symmetry conditions.

5. Concluding remarks

If the implicit gradient elasticity model in Eq. (2), named the 3-PG-Model, is recognized as a particular case of micromorphic (micro-strain) elasticity, a free energy and associated response functions and boundary conditions can be assigned. Part *I* adopts this conceptual point of view for the 3-PG-Model and provides the reduced form of the governing equations and boundary conditions for plane strain problems. This includes, among others, elasticity laws for classical and nonclassical stresses as well as classical and nonclassical equilibrium equations and compatibility conditions. It also supplies the required symmetry conditions for asymptotic solutions of Mode-I and Mode-II crack problems. A detailed discussion of such analytical solutions is given in Part *II*.

A. Appendix

This section provides the component representations with respect to cylindrical coordinates of some space derivatives of a second-order tensor $\boldsymbol{A}$ and a third-order tensor $\boldsymbol{\mathcal{A}}$. It is easy to find component representations for $\nabla\boldsymbol{A}$ and $\operatorname{div}\boldsymbol{A}$ in textbooks, whereas it may be harder to find such representations for $\Delta\boldsymbol{A}$ and $\operatorname{div}\boldsymbol{\mathcal{A}}$. But one can calculate them with the help of the relations given below.

A.1 Cylindrical coordinates

We denote by $\{\theta^i\}$ the cylindrical coordinate system with $\theta^1 = r, \theta^2 = \varphi$ and $\theta^3 = z$. The covariant basis induced by $\{\theta^i\}$ is denoted by $\{\boldsymbol{g}_i\}$ where $\boldsymbol{g}_1 = (\cos\varphi)\boldsymbol{e}_1 + (\sin\varphi)\boldsymbol{e}_2, \boldsymbol{g}_2 = -r(\sin\varphi)\boldsymbol{e}_1 + r(\cos\varphi)\boldsymbol{e}_2$ and $\boldsymbol{g}_3 = \boldsymbol{e}_3$. The contravariant basis is denoted by $\{\boldsymbol{g}^i\}, \boldsymbol{g}^i = g^{ij}\boldsymbol{g}_j$, where g^{ij} are the contravariant metric coefficients. The corresponding covariant metric coefficients are g_{ij}. All values of g_{ij} and g^{ij} are vanishing except for the values $g_{11} = g_{33} = g^{11} = g^{33} = 1, g_{22} = r^2$ and $g^{22} = \frac{1}{r^2}$. Moreover, all values of the related Christoffel symbols Γ^k_{ij} also vanish except for the values $\Gamma^1_{22} = -r$ and $\Gamma^2_{12} = \Gamma^2_{21} = \frac{1}{r}$. Physical components are referred to the orthonormal basis $\{\boldsymbol{e}_{<i>}\}$ where $\boldsymbol{e}_{<1>} \equiv \boldsymbol{e}_r = \boldsymbol{g}_1, \boldsymbol{e}_{<2>} \equiv \boldsymbol{e}_\varphi = \frac{1}{r}\boldsymbol{g}_2$ and $\boldsymbol{e}_{<3>} \equiv \boldsymbol{e}_z = \boldsymbol{g}_3$. According to Section 2, the physical components of a second-order tensor $\boldsymbol{A}$ are denoted by $A_{\alpha\beta}$ and notations of the form $A_{<11>} \equiv A_{rr}, A_{<12>} \equiv A_{r\varphi}$, ... apply. Similar notations hold also for any tensor, especially for the third-order tensor $\boldsymbol{\mathcal{A}}$. Finally, the nabla operator ∇ obeys the representation $\nabla = \frac{\partial}{\partial\theta^i}\boldsymbol{g}^i$.

A.2 The gradient of a symmetric second-order tensor

In the case of a second-order tensor $\boldsymbol{A} = A^{ij}\boldsymbol{g}_i \otimes \boldsymbol{g}_j$, we have (cf. Section 2)

$$\operatorname{grad}\boldsymbol{A} \equiv \nabla\boldsymbol{A} := \nabla \otimes \boldsymbol{A} = \boldsymbol{g}^i \otimes \partial_{\theta^i}\boldsymbol{A} = (\nabla\boldsymbol{A})_i{}^{jk}\boldsymbol{g}^i \otimes \boldsymbol{g}_j \otimes \boldsymbol{g}_k, \tag{A1}$$

where $(\nabla\boldsymbol{A})_i{}^{jk}$ is the covariant derivative of the components A^{jk},

$$(\nabla\boldsymbol{A})_i{}^{jk} = A^{jk}\Big|_i = \partial_{\theta^i}A^{jk} + \Gamma^j_{im}A^{mk} + \Gamma^k_{im}A^{jm}. \tag{A2}$$

When $\boldsymbol{A}$ is symmetric, $\boldsymbol{A} = \boldsymbol{A}^{(s)}$, we conclude from Eq. (A2), that

$$(\nabla\boldsymbol{A})_i{}^{jk} = (\nabla\boldsymbol{A})_i{}^{(jk)}. \tag{A3}$$

This symmetry also applies with respect to physical components,

$$(\nabla\boldsymbol{A})_{\alpha\beta\gamma} = (\nabla\boldsymbol{A})_{\alpha(\beta\gamma)}. \tag{A4}$$

It can be seen that Eq. (A2) furnishes the following physical components of $\nabla\boldsymbol{A}$:

$$(\nabla\boldsymbol{A})_{rrr} = \partial_r A_{rr}, \quad (\nabla\boldsymbol{A})_{r\varphi\varphi} = \partial_r A_{\varphi\varphi}, \quad (\nabla\boldsymbol{A})_{rzz} = \partial_r A_{zz}, \tag{A5}$$

$$(\nabla\boldsymbol{A})_{rr\varphi} = \partial_r A_{r\varphi}, \quad (\nabla\boldsymbol{A})_{rrz} = \partial_r A_{rz}, \quad (\nabla\boldsymbol{A})_{r\varphi z} = \partial_r A_{\varphi z}, \tag{A6}$$

$$(\nabla\boldsymbol{A})_{\varphi rr} = \frac{1}{r}\left(\partial_\varphi A_{rr} - 2A_{\varphi r}\right), \tag{A7}$$

$$(\nabla\boldsymbol{A})_{\varphi\varphi\varphi} = \frac{1}{r}\left(\partial_\varphi A_{\varphi\varphi} + 2A_{\varphi r}\right), \tag{A8}$$

$$(\nabla\boldsymbol{A})_{\varphi zz} = \frac{1}{r}\partial_\varphi A_{zz}, \tag{A9}$$

$$(\nabla\boldsymbol{A})_{\varphi r\varphi} = \frac{1}{r}\left(\partial_\varphi A_{r\varphi} + A_{rr} - A_{\varphi\varphi}\right), \tag{A10}$$

$$(\nabla\boldsymbol{A})_{\varphi\varphi z} = \frac{1}{r}\left(\partial_\varphi A_{\varphi z} - A_{rz}\right), \tag{A11}$$

$$(\nabla\boldsymbol{A})_{\varphi rz} = \frac{1}{r}\left(\partial_\varphi A_{rz} - A_{\varphi z}\right), \tag{A12}$$

$$(\nabla\boldsymbol{A})_{z\alpha\beta} = 0. \tag{A13}$$

A.3 The Laplacian of a symmetric second-order tensor

The Laplacian of a second-order tensor $\boldsymbol{A}$ is given by (cf. Section 2)

$$\Delta \boldsymbol{A} := \operatorname{div} \operatorname{grad} \boldsymbol{A} = \nabla \cdot (\nabla \boldsymbol{A}) = \boldsymbol{g}^m \cdot \partial_{\theta^m} (\nabla \boldsymbol{A}) \\ = \boldsymbol{g}^m \cdot \partial_{\theta^m} \left(A^{jk}|_i \boldsymbol{g}^i \otimes \boldsymbol{g}_j \otimes \boldsymbol{g}_k \right). \quad \text{(A14)}$$

This may be written as

$$\Delta \boldsymbol{A} = (\Delta \boldsymbol{A})^{ij} \boldsymbol{g}_i \otimes \boldsymbol{g}_j, \quad \text{(A15)}$$

where

$$(\Delta \boldsymbol{A})^{ij} = A^{ij}\Big|_{km} g^{km}, \quad \text{(A16)}$$

and $A^{ij}|_{km}$ is the second covariant derivative of the components A^{ij},

$$A^{ij}|_{km} = \left(A^{ij}{}_k\right)\Big|_m = \partial_{\theta^m} A^{ij}|_k + \Gamma^i_{ml} A^{lj}\big|_k + \Gamma^j_{ml} A^{il}|_k - \Gamma^l_{km} A^{ij}\big|_l. \quad \text{(A17)}$$

We can calculate the physical components $(\Delta \boldsymbol{A})_{<ij>}$ from Eqs. (A16) and (A17). For the case that $\boldsymbol{A}$ is symmetric, $\boldsymbol{A} = \boldsymbol{A}^{(s)}$, we can derive, after lengthy algebraic manipulations, that

$$(\Delta \boldsymbol{A})_{rr} = \partial_{rr} A_{rr} + \frac{1}{r^2} \partial_{\varphi\varphi} A_{rr} + \partial_{zz} A_{rr} + \frac{1}{r} \partial_r A_{rr} - \frac{4}{r^2} \partial_\varphi A_{r\varphi} \\ - \frac{2}{r^2} A_{rr} + \frac{2}{r^2} A_{\varphi\varphi}, \quad \text{(A18)}$$

$$(\Delta \boldsymbol{A})_{\varphi\varphi} = \partial_{rr} A_{\varphi\varphi} + \frac{1}{r^2} \partial_{\varphi\varphi} A_{\varphi\varphi} + \partial_{zz} A_{\varphi\varphi} + \frac{1}{r} \partial_r A_{\varphi\varphi} + \frac{4}{r^2} \partial_\varphi A_{r\varphi} \\ + \frac{2}{r^2} A_{rr} - \frac{2}{r^2} A_{\varphi\varphi}, \quad \text{(A19)}$$

$$(\Delta \boldsymbol{A})_{zz} = \partial_{rr} A_{zz} + \frac{1}{r^2} \partial_{\varphi\varphi} A_{zz} + \partial_{zz} A_{zz} + \frac{1}{r} \partial_r A_{zz}, \quad \text{(A20)}$$

$$(\Delta \boldsymbol{A})_{r\varphi} = \partial_{rr} A_{r\varphi} + \frac{1}{r^2} \partial_{\varphi\varphi} A_{r\varphi} + \partial_{zz} A_{r\varphi} + \frac{1}{r} \partial_r A_{r\varphi} + \frac{2}{r^2} \partial_\varphi A_{rr} \\ - \frac{2}{r^2} \partial_\varphi A_{\varphi\varphi} - \frac{4}{r^2} A_{r\varphi}, \quad \text{(A21)}$$

$$(\Delta \boldsymbol{A})_{rz} = \partial_{rr} A_{rz} + \frac{1}{r^2} \partial_{\varphi\varphi} A_{rz} + \partial_{zz} A_{rz} + \frac{1}{r} \partial_r A_{rz} - \frac{2}{r^2} \partial_\varphi A_{\varphi z} - \frac{1}{r^2} A_{rz}, \quad \text{(A22)}$$

$$(\Delta \boldsymbol{A})_{\varphi z} = \partial_{rr} A_{\varphi z} + \frac{1}{r^2} \partial_{\varphi\varphi} A_{\varphi z} + \partial_{zz} A_{\varphi z} + \frac{1}{r} \partial_r A_{\varphi z} + \frac{2}{r^2} \partial_\varphi A_{rz} - \frac{1}{r^2} A_{\varphi z}. \quad \text{(A23)}$$

A.4. The divergence of a third-order tensor

Let $\mathcal{A} = \mathcal{A}^{ijk} \boldsymbol{g}_i \otimes \boldsymbol{g}_j \otimes \boldsymbol{g}_k$ be a third-order tensor. Then (cf. Section 2)

$$\operatorname{div} \mathcal{A} := \nabla \cdot \mathcal{A} = \boldsymbol{g}^i \cdot \partial_{\theta^i} \mathcal{A} = (\operatorname{div} \mathcal{A})^{jk} \boldsymbol{g}_j \otimes \boldsymbol{g}_k, \quad \text{(A24)}$$

where

$$(\operatorname{div}\boldsymbol{\mathcal{A}})^{jk} = \mathcal{A}^{ijk}\Big|_{i}, \tag{A25}$$

and $\mathcal{A}^{ijk}\Big|_{m}$ is the covariant derivative of the components $\mathcal{A}^{ijk}$. If the symmetry condition $\mathcal{A}^{ijk} \equiv \mathcal{A}^{i(jk)}$ holds, we can establish, after lengthy algebraic manipulations, the following results for the physical components of $\operatorname{div}\boldsymbol{\mathcal{A}}$:

$$(\operatorname{div}\boldsymbol{\mathcal{A}})_{rr} = \partial_r A_{rrr} + \frac{1}{r}\,\partial_\varphi A_{\varphi rr} + \partial_z A_{zrr} + \frac{1}{r}\left(A_{rrr} - 2A_{\varphi r\varphi}\right), \tag{A26}$$

$$(\operatorname{div}\boldsymbol{\mathcal{A}})_{\varphi\varphi} = \partial_r A_{r\varphi\varphi} + \frac{1}{r}\,\partial_\varphi A_{\varphi\varphi\varphi} + \partial_z A_{z\varphi\varphi} + \frac{1}{r}\left(A_{r\varphi\varphi} + 2A_{\varphi r\varphi}\right), \tag{A27}$$

$$(\operatorname{div}\boldsymbol{\mathcal{A}})_{zz} = \partial_r A_{rzz} + \frac{1}{r}\,\partial_\varphi A_{\varphi zz} + \partial_z A_{zzz} + \frac{1}{r}\,A_{rzz}, \tag{A28}$$

$$(\operatorname{div}\boldsymbol{\mathcal{A}})_{r\varphi} = \partial_r A_{rr\varphi} + \frac{1}{r}\,\partial_\varphi A_{\varphi r\varphi} + \partial_z A_{zr\varphi} + \frac{1}{r}\left(A_{rr\varphi} - A_{\varphi\varphi\varphi} + A_{\varphi rr}\right), \tag{A29}$$

$$(\operatorname{div}\boldsymbol{\mathcal{A}})_{rz} = \partial_r A_{rrz} + \frac{1}{r}\,\partial_\varphi A_{\varphi rz} + \partial_z A_{zrz} + \frac{1}{r}\left(A_{rrz} - A_{\varphi\varphi z}\right), \tag{A30}$$

$$(\operatorname{div}\boldsymbol{\mathcal{A}})_{\varphi z} = \partial_r A_{r\varphi z} + \frac{1}{r}\,\partial_\varphi A_{\varphi\varphi z} + \partial_z A_{z\varphi z} + \frac{1}{r}\left(A_{r\varphi z} + A_{\varphi rz}\right). \tag{A31}$$

References

[1] Altan BS, Aifantis EC. On some aspects in the special theory of gradient elasticity. Journal of the Mechanical Behavior of Materials. 1997;**8**:231-282. DOI: 10.1007/s00161-014-0406-1

[2] Georgiadis HG. The mode III crack problem in microstructured solids governed by dipolar gradient elasticity: Static and dynamic analysis. Journal of Applied Mechanics. 2003;**70**:517-530

[3] Mindlin RD. Micro-structure in linear elasticity. Archive for Rational Mechanics and Analysis. 1964;**16**:51-78

[4] Mindlin RD, Eshel NN. On first strain-gradient theories in linear elasticity. International Journal of Solids and Structures. 1968;**4**:109-124

[5] Broese C, Tsakmakis C, Beskos DE. Gradient elasticity based on Laplacians of stress and strain. Journal of Elasticity. 2018;**131**:39-74

[6] Gutkin M, Aifantis EC. Dislocations in the theory of gradient elasticity. Scripta Materialia. 1999;**40**:559-566

[7] Askes H, Gutiérrez MA. Implicit gradient elasticity. International Journal for Numerical Methods in Engineering. 2006;**67**:400-416

[8] Eringen AC, Suhubi ES. Nonlinear theory of simple micro-elastic solids–I. International Journal of Engineering Science. 1964;**2**:189-203

[9] Eringen AC. Theory of micropolar elasticity. In: Fracture – An Advanced Treatise. Vol. 1. Microscopic and Macroscopic Fundamentals. New York: Academic Press; 1968. pp. 621-729

[10] Forest S, Sievert R. Nonlinear microstrain theories. International Journal of Solids and Structures. 2006; **43**(24):7224-7245

[11] Alber H-D, Hutter K, Tsakmakis C. Nonconventional thermodynamics, indeterminate couple stress elasticity and heat conduction. Continuum Mechanics and Thermodynamics. 2014. DOI: 10.1007/s00161-014-0406-1

[12] Anderson TL. Fracture Mechanics - Fundamentals and Applications. CRC Press; 1991

[13] Malvern LE. Introduction to the Mechanics of a Continuous Medium. Englewood Cliffs, New Jersey: Prentice Hall, Inc.; 1969

[14] Hellan K. Introduction to Fracture Mechanics. New York: McGraw-Hill; 1984

Chapter 3

Mode-I and Mode-II Crack Tip Fields in Implicit Gradient Elasticity Based on Laplacians of Stress and Strain. Part II: Asymptotic Solutions

Abstract

We develop asymptotic solutions for near-tip fields of Mode-I and Mode-II crack problems and for model responses reflected by implicit gradient elasticity. Especially, a model of gradient elasticity is considered, which is based on Laplacians of stress and strain and turns out to be derivable as a particular case of micromorphic (microstrain) elasticity. While the governing model equations of the crack problems are developed in Part I, the present paper addresses analytical solutions for near-tip fields by using asymptotic expansions of Williams' type. It is shown that for the assumptions made in Part I, the model does not eliminiate the well-known singularities of classical elasticity. This is in contrast to conclusions made elsewhere, which rely upon different assumptions. However, there are significant differences in comparison to classical elasticity, which are discussed in the paper. For instance, in the case of Mode-II loading conditions, the leading terms of the asymptotic solution for the components of the double stress exhibit the remarkable property that they include two stress intensity factors.

Keywords: implicit gradient elasticity, mode-I and mode-II crack problems, analytical solutions, asymptotic expansions of Williams' type, near-tip fields, order of singularity, stress intensity factors

1. Introduction

The 3-PG-Model, discussed in Part I, is a simple model of implicit gradient elasticity based on Laplacians of stress and strain and has been introduced by Gutkin and Aifantis [1]. It can be derived as a particular case of micromorphic (microstrain) elasticity (see, e. g., Forest and Sievert [2]), so that a free energy function and required boundary conditions are formulated rigorously. In the present paper, we are looking for near-tip asymptotic field solutions for Mode-I and Mode-II crack problems, in the context of plane strain states. The asymptotic solutions are obtained by using expansions of Williams' type (see Williams [3]).

For the assumptions made in Part I, it is found that both, conventional stress and conventional strain, are singular. This holds also for the nonconventional stress, the so-called double stress. All singular fields have an order of singularity $r^{-\frac{1}{2}}$. In particular, the leading terms of the asymptotic solutions of the conventional stress are exactly the same as in classical elasticity. Nevertheless, the results are quite interesting, since the two leading terms of the asymptotic solution of the macrostrain are different from the corresponding terms of classical elasticity, and since the form of the asymptotic solution of the double stress exhibits a remarkable feature. To be more specific, the leading term of the asymptotic solution of the double stress includes two stress intensity factors, which are independent of each other. This reflects, from a theoretical point of view, differences in the structure of the asymptotic solutions in comparison to classical elasticity as well as micropolar elasticity, where only one stress intensity factor is present in the solutions of Mode-II crack problems.

There are various works addressing singularities in the field variables. Among others, we mention for couple-stress elasticity the works of Muki and Sternberg [4], Sternberg and Muki [5], Bogy and Sternberg [6, 7], Xia and Hutchinson [8], Huang et al. [9–11] and Zhang et al. [12]. For micropolar elasticity the works of Paul and Sridharan [13], Chen et al. [14], Diegele et al. [15] and for gradient elasticity the works of Altan and Aifantis [16, 17], Ru and Aifantis [18], Unger and Aifantis [19–21], Chen et al. [22], Mousavi and Lazar [23], Shi et al. [24, 25], Vardoulakis et al. [26], Karlis et al. [27, 28], Georgiadis [29], Askes and Aifantis [30] and Gutkin and Aifantis [1] are to be mentioned. The latter is an interesting work and proves that use of the 3-PG-Model eliminates singularities from the"elastic stresses of defects" (see also Askes and Aifantis [30] as well as Aifantis [31]). This finding is in contrast to the conclusions of the present paper, but it should be emphasized that the form of the assumed boundary conditions in Gutkin and Aifantis [1] is different from the form assumed here.

The scope of the paper is organized as follows: Mode-I and Mode-II crack problems are considered in the setting of plane strain problems. For the 3-PG-Model, the reduced governing equations for plane strain states have been derived in Part I and are summarized in Section 2. Section 3 provides asymptotic solutions for the near-tip fields by starting from asymptotic expansions of the macrodisplacement and the microdeformation. An alternative and equivalent aproach, starting from asymptotic expansions of the stresses, is sketched in Section 4. The developed asymptotic solutions are summarized and discussed in Section 5. Finally, the paper closes with some conclusions in Section 6.

Throughout the paper, use is made of the notation introduced in Part I.

2. Summary of the governing equations for plane strain problems

Following equations of Part I will be employed to establish asymptotic solutions of the crack tip fields.

Free energy function (see section "The 3-PG-Model as particular case of micro-strain elasticity" in Part I)

$$\psi = \frac{1}{2}\,\varepsilon_{\alpha\beta}\mathbb{C}_{\alpha\beta\rho\zeta}\,\varepsilon_{\rho\zeta} + \frac{1}{2}\frac{c_2 - c_1}{c_1}\,\gamma_{\alpha\beta}\mathbb{C}_{\alpha\beta\rho\zeta}\,\gamma_{\rho\zeta} + \frac{1}{2}\,(c_2 - c_1)\,k_{\alpha\beta\gamma}\,\mathbb{C}_{\beta\gamma\rho\zeta}\,k_{\alpha\rho\zeta}. \tag{1}$$

Elasticity law for $\mathbf{\Sigma}$ (see section 3.1 "The 3-PG-Model as particular case of micro-strain elasticity" in Part I)

$$\Sigma_{\alpha\beta} = \frac{c_2}{c_1}\,\mathbb{C}_{\alpha\beta\gamma\rho}\,\varepsilon_{\gamma\rho} - \frac{c_2 - c_1}{c_1}\,\mathbb{C}_{\alpha\beta\gamma\rho}\,\Psi_{\gamma\rho} \tag{2}$$

or inversely

$$\varepsilon_{rr} = \frac{c_1}{2\mu c_2}\left[\Sigma_{rr} - \nu\left(\Sigma_{rr} + \Sigma_{\varphi\varphi}\right)\right] + \frac{c_2 - c_1}{c_1}\Psi_{rr}, \tag{3}$$

$$\varepsilon_{\varphi\varphi} = \frac{c_1}{2\mu c_2}\left[\Sigma_{\varphi\varphi} - \nu\left(\Sigma_{rr} + \Sigma_{\varphi\varphi}\right)\right] + \frac{c_2 - c_1}{c_1}\Psi_{\varphi\varphi}, \tag{4}$$

$$\varepsilon_{r\varphi} = \frac{c_1}{2\mu c_2}\Sigma_{r\varphi} + \frac{c_2 - c_1}{c_1}\Psi_{r\varphi}. \tag{5}$$

Elasticity law for $\boldsymbol{\sigma}$ (see section 3.1 "The 3-PG-Model as particular case of micro-strain elasticity" in Part I)

$$\sigma_{\alpha\beta} = \frac{c_2 - c_1}{c_1}\mathbb{C}_{\alpha\beta\rho\zeta}\left(\varepsilon_{\rho\zeta} - \Psi_{\rho\zeta}\right). \tag{6}$$

Elasticity law for $\boldsymbol{\mu}$ (see section 4.5.1 "Elasticity law for double stress" in Part I)

$$\mu_{rrr} = (c_2 - c_1)\left[(\lambda + 2\mu)\partial_r\Psi_{rr} + \lambda\partial_r\Psi_{\varphi\varphi}\right], \tag{7}$$

$$\mu_{r\varphi\varphi} = (c_2 - c_1)\left[(\lambda + 2\mu)\partial_r\Psi_{\varphi\varphi} + \lambda\partial_r\Psi_{rr}\right], \tag{8}$$

$$\mu_{rzz} = (c_2 - c_1)\lambda\partial_r\left(\Psi_{rr} + \Psi_{\varphi\varphi}\right) = \nu\left(\mu_{rrr} + \mu_{r\varphi\varphi}\right), \tag{9}$$

$$\mu_{rr\varphi} = (c_2 - c_1)2\mu\partial_r\Psi_{r\varphi}, \tag{10}$$

$$\mu_{\varphi rr} = \frac{c_2 - c_1}{r}\left[2\mu\left(\partial_\varphi\Psi_{rr} - 2\Psi_{r\varphi}\right) + \lambda\partial_\varphi\left(\Psi_{rr} + \Psi_{\varphi\varphi}\right)\right], \tag{11}$$

$$\mu_{\varphi\varphi\varphi} = \frac{c_2 - c_1}{r}\left[2\mu\left(\partial_\varphi\Psi_{\varphi\varphi} + 2\Psi_{r\varphi}\right) + \lambda\partial_\varphi\left(\Psi_{rr} + \Psi_{\varphi\varphi}\right)\right], \tag{12}$$

$$\mu_{\varphi zz} = \frac{c_2 - c_1}{r}\lambda\partial_\varphi\left(\Psi_{rr} + \Psi_{\varphi\varphi}\right) = \nu\left(\mu_{\varphi rr} + \mu_{\varphi\varphi\varphi}\right), \tag{13}$$

$$\mu_{\varphi r\varphi} = \frac{c_2 - c_1}{r}2\mu\left(\partial_\varphi\Psi_{r\varphi} + \Psi_{rr} - \Psi_{\varphi\varphi}\right), \tag{14}$$

$$\mu_{rrz} = \mu_{r\varphi z} = \mu_{\varphi rz} = \mu_{\varphi\varphi z} = 0, \tag{15}$$

$$\mu_{z\alpha\beta} = 0, \tag{16}$$

or inversely

$$(\nabla\Psi)_{rrr} = \frac{1}{(c_2 - c_1)E}\left[(1 - \nu^2)\mu_{rrr} - \nu(1 + \nu)\mu_{r\varphi\varphi}\right], \tag{17}$$

$$(\nabla\Psi)_{r\varphi\varphi} = \frac{1}{(c_2 - c_1)E}\left[(1 - \nu^2)\mu_{r\varphi\varphi} - \nu(1 + \nu)\mu_{rrr}\right], \tag{18}$$

$$(\nabla\Psi)_{rr\varphi} = \frac{1 + \nu}{(c_2 - c_1)E}\mu_{rr\varphi}, \tag{19}$$

$$(\nabla\Psi)_{\varphi rr} = \frac{1}{(c_2 - c_1)E}\left[(1 - \nu^2)\mu_{\varphi rr} - \nu(1 + \nu)\mu_{\varphi\varphi\varphi}\right], \tag{20}$$

$$(\nabla\Psi)_{\varphi\varphi\varphi} = \frac{1}{(c_2 - c_1)E}\left[(1 - \nu^2)\mu_{\varphi\varphi\varphi} - \nu(1 + \nu)\mu_{\varphi rr}\right], \tag{21}$$

$$(\nabla\Psi)_{\varphi r\varphi} = \frac{1 + \nu}{(c_2 - c_1)E}\mu_{\varphi r\varphi}. \tag{22}$$

Material parameters (see section 2 "Preliminaries—Notation" in Part I)

$$\nu = \frac{\lambda}{2(\lambda+\mu)}, \quad E = 2\mu(1+\nu). \tag{23}$$

Strain components (see section 4.1 "Kinematics" in Part I)

$$\varepsilon_{rr} = \partial_r u_r, \quad \varepsilon_{\varphi\varphi} = \frac{1}{r}\left(u_r + \partial_\varphi u_\varphi\right), \quad \varepsilon_{r\varphi} = \frac{1}{2}\left(\frac{1}{r}\,\partial_\varphi u_r + \partial_r u_\varphi - \frac{1}{r}\,u_\varphi\right). \tag{24}$$

Microdeformation components (see section 4.1 "Kinematics" of Part I)

$$(\nabla\boldsymbol{\Psi})_{rrr} = \partial_r \boldsymbol{\Psi}_{rr}, \quad (\nabla\boldsymbol{\Psi})_{r\varphi\varphi} = \partial_r \boldsymbol{\Psi}_{\varphi\varphi}, \quad (\nabla\boldsymbol{\Psi})_{rr\varphi} = \partial_r \Psi_{r\varphi}, \tag{25}$$

$$(\nabla\boldsymbol{\Psi})_{\varphi rr} = \frac{1}{r}\left(\partial_\varphi \Psi_{rr} - 2\Psi_{r\varphi}\right), \tag{26}$$

$$(\nabla\boldsymbol{\Psi})_{\varphi\varphi\varphi} = \frac{1}{r}\left(\partial_\varphi \Psi_{\varphi\varphi} + 2\Psi_{r\varphi}\right), \tag{27}$$

$$(\nabla\boldsymbol{\Psi})_{\varphi r\varphi} = \frac{1}{r}\left(\partial_\varphi \Psi_{r\varphi} + \Psi_{rr} - \Psi_{\varphi\varphi}\right), \tag{28}$$

$$(\nabla\boldsymbol{\Psi})_{\alpha\beta z} = (\nabla\boldsymbol{\Psi})_{z\alpha\beta} = 0. \tag{29}$$

Classical equilibrium equations (see section 4.2 "Cauchy stress—Classical equilibrium equations" in Part I)

$$\partial_r \Sigma_{rr} + \frac{1}{r}\,\partial_\varphi \Sigma_{r\varphi} + \frac{1}{r}\left(\Sigma_{rr} - \Sigma_{\varphi\varphi}\right) = 0, \tag{30}$$

$$\partial_r \Sigma_{r\varphi} + \frac{1}{r}\,\partial_\varphi \Sigma_{\varphi\varphi} + \frac{2}{r}\,\Sigma_{r\varphi} = 0. \tag{31}$$

Nonclassical equilibrium equations (see section 4.5.2 "Nonclassical equilibrium conditions" in Part I)

$$\partial_r \mu_{rrr} + \frac{1}{r}\,\partial_\varphi \mu_{\varphi rr} + \frac{1}{r}\left(\mu_{rrr} - 2\mu_{\varphi r\varphi}\right) + \sigma_{rr} = 0, \tag{32}$$

$$\partial_r \mu_{r\varphi\varphi} + \frac{1}{r}\,\partial_\varphi \mu_{\varphi\varphi\varphi} + \frac{1}{r}\left(\mu_{r\varphi\varphi} + 2\mu_{\varphi r\varphi}\right) + \sigma_{\varphi\varphi} = 0, \tag{33}$$

$$\partial_r \mu_{rzz} + \frac{1}{r}\,\partial_\varphi \mu_{\varphi zz} + \frac{1}{r}\,\mu_{rzz} + \sigma_{zz} = 0, \tag{34}$$

$$\partial_r \mu_{rr\varphi} + \frac{1}{r}\,\partial_\varphi \mu_{\varphi r\varphi} + \frac{1}{r}\left(\mu_{rr\varphi} - \mu_{\varphi\varphi\varphi} + \mu_{\varphi rr}\right) + \sigma_{r\varphi} = 0. \tag{35}$$

Field equations for Ψ (see section 4.4 "Field equations for Ψ" in Part I)

$$\begin{aligned} &\partial_{rr}\Psi_{rr} + \frac{1}{r^2}\,\partial_{\varphi\varphi}\Psi_{rr} + \frac{1}{r}\,\partial_r \Psi_{rr} - \frac{4}{r^2}\,\partial_\varphi \Psi_{r\varphi} - \left(\frac{2}{r^2} + \frac{1}{c_2}\right)\Psi_{rr} + \frac{2}{r^2}\,\Psi_{\varphi\varphi} \\ &+ \frac{1-\nu}{2\mu c_2}\,\Sigma_{rr} - \frac{\nu}{2\mu c_2}\,\Sigma_{\varphi\varphi} = 0, \end{aligned} \tag{36}$$

$$\begin{aligned} &\partial_{rr}\Psi_{\varphi\varphi} + \frac{1}{r^2}\,\partial_{\varphi\varphi}\Psi_{\varphi\varphi} + \frac{1}{r}\,\partial_r \Psi_{\varphi\varphi} + \frac{4}{r^2}\,\partial_\varphi \Psi_{r\varphi} + \frac{2}{r^2}\,\Psi_{rr} - \left(\frac{2}{r^2} + \frac{1}{c_2}\right)\Psi_{\varphi\varphi} \\ &+ \frac{1-\nu}{2\mu c_2}\,\Sigma_{\varphi\varphi} - \frac{\nu}{2\mu c_2}\,\Sigma_{rr} = 0, \end{aligned} \tag{37}$$

$$\begin{aligned}\partial_{rr}\Psi_{r\varphi}+\frac{1}{r^2}\,\partial_{\varphi\varphi}\Psi_{r\varphi}+\frac{1}{r}\,\partial_r\Psi_{r\varphi}+\frac{2}{r^2}\,\partial_\varphi\Psi_{rr}-\frac{2}{r^2}\,\partial_\varphi\Psi_{\varphi\varphi}-\left(\frac{4}{r^2}+\frac{1}{c_2}\right)\Psi_{r\varphi}\\+\frac{1}{2\mu c_2}\,\Sigma_{r\varphi}=0.\end{aligned}\tag{38}$$

Classical compatibility condition (see section 4.3 "Classical compatibility condition" in Part I)

$$\chi_1(\Psi_{\alpha\beta})+\chi_2(\Sigma_{\alpha\beta})=0,\tag{39}$$

$$\begin{aligned}\chi_1(\Psi_{\alpha\beta}):=\frac{c_2-c_1}{c_2}\Big[\partial_{rr}\Psi_{\varphi\varphi}-\frac{2}{r}\,\partial_{r\varphi}\Psi_{r\varphi}+\frac{1}{r^2}\,\partial_{\varphi\varphi}\Psi_{rr}\\-\frac{1}{r}\,\partial_r\Psi_{rr}+\frac{2}{r}\,\partial_r\Psi_{\varphi\varphi}-\frac{2}{r^2}\,\partial_\varphi\Psi_{r\varphi}\Big],\end{aligned}\tag{40}$$

$$\begin{aligned}\chi_2(\Sigma_{\alpha\beta}):=\frac{(1-\nu)\,c_1}{2\mu c_2}\Big[\partial_{rr}(\Sigma_{rr}+\Sigma_{\varphi\varphi})+\frac{1}{r^2}\,\partial_{\varphi\varphi}(\Sigma_{rr}+\Sigma_{\varphi\varphi})\\+\frac{1}{r}\,\partial_r(\Sigma_{rr}+\Sigma_{\varphi\varphi})\Big].\end{aligned}\tag{41}$$

Nonclassical compatibility conditions (see section 4.6 "Nonclassical compatibility conditions" in Part I)

$$\partial_\varphi\mu_{rr\varphi}-\mu_{\varphi r\varphi}-r\,\partial_r\mu_{\varphi r\varphi}+\mu_{rrr}-\mu_{r\varphi\varphi}=0,\tag{42}$$

$$\partial_\varphi\mu_{r\varphi\varphi}+\partial_\varphi\mu_{rrr}-\mu_{\varphi\varphi\varphi}-\mu_{\varphi rr}-r\,\partial_r\mu_{\varphi\varphi\varphi}-r\,\partial_r\mu_{\varphi rr}=0,\tag{43}$$

$$\partial_\varphi\mu_{r\varphi\varphi}-\partial_\varphi\mu_{rrr}-\mu_{\varphi\varphi\varphi}+\mu_{\varphi rr}-r\,\partial_r\mu_{\varphi\varphi\varphi}+r\,\partial_r\mu_{\varphi rr}+4\mu_{rr\varphi}=0.\tag{44}$$

Classical boundary conditions (see section 4.7 "Boundary conditions" in Part I)

$$\left[\Sigma_{r\varphi}\right]_{\varphi=\pm\pi}=0,\tag{45}$$

$$\left[\Sigma_{\varphi\varphi}\right]_{\varphi=\pm\pi}=0.\tag{46}$$

Nonclassical boundary conditions (see section 4.7 "Boundary conditions" in Part I)

$$\left[\mu_{\varphi rr}\right]_{\varphi=\pm\pi}=\left[\mu_{\varphi\varphi\varphi}\right]_{\varphi=\pm\pi}=\left[\mu_{\varphi r\varphi}\right]_{\varphi=\pm\pi}=0,\tag{47}$$

or equivalently

$$\left[\partial_\varphi\Psi_{rr}-2\Psi_{r\varphi}\right]_{\varphi=\pm\pi}=0,\tag{48}$$

$$\left[\partial_\varphi\Psi_{\varphi\varphi}+2\Psi_{r\varphi}\right]_{\varphi=\pm\pi}=0,\tag{49}$$

$$\left[\partial_\varphi\Psi_{r\varphi}+\Psi_{rr}-\Psi_{\varphi\varphi}\right]_{\varphi=\pm\pi}=0.\tag{50}$$

Symmetry conditions—Mode-I (see section 4.8 "Symmetry conditions" in Part I)

$$\Sigma_{rr}(r,\varphi)=\Sigma_{rr}(r,-\varphi),\quad \Sigma_{\varphi\varphi}(r,\varphi)=\Sigma_{\varphi\varphi}(r,-\varphi),\tag{51}$$

$$\Sigma_{r\varphi}(r,\varphi)=-\Sigma_{r\varphi}(r,-\varphi),\tag{52}$$

$$\Psi_{rr}(r,\varphi) = \Psi_{rr}(r,-\varphi), \quad \Psi_{\varphi\varphi}(r,\varphi) = \Psi_{\varphi\varphi}(r,-\varphi), \tag{53}$$

$$\Psi_{r\varphi}(r,\varphi) = -\Psi_{r\varphi}(r,-\varphi), \tag{54}$$

$$\mu_{rrr}(r,\varphi) = \mu_{rrr}(r,-\varphi), \quad \mu_{\varphi rr}(r,\varphi) = -\mu_{\varphi rr}(r,-\varphi), \tag{55}$$

$$\mu_{r\varphi\varphi}(r,\varphi) = \mu_{r\varphi\varphi}(r,-\varphi), \quad \mu_{\varphi\varphi\varphi}(r,\varphi) = -\mu_{\varphi\varphi\varphi}(r,-\varphi), \tag{56}$$

$$\mu_{rzz}(r,\varphi) = \mu_{rzz}(r,-\varphi), \quad \mu_{\varphi zz}(r,\varphi) = -\mu_{\varphi zz}(r,-\varphi), \tag{57}$$

$$\mu_{rr\varphi}(r,\varphi) = \mu_{rr\varphi}(r,-\varphi), \quad \mu_{\varphi r\varphi}(r,\varphi) = -\mu_{\varphi r\varphi}(r,-\varphi). \tag{58}$$

Symmetry conditions—Mode-II (see section 4.8 "Symmetry conditions" in Part I)

$$\Sigma_{rr}(r,\varphi) = -\Sigma_{rr}(r,-\varphi), \quad \Sigma_{\varphi\varphi}(r,\varphi) = -\Sigma_{\varphi\varphi}(r,-\varphi), \tag{59}$$

$$\Sigma_{r\varphi}(r,\varphi) = \Sigma_{r\varphi}(r,-\varphi), \tag{60}$$

$$\Psi_{rr}(r,\varphi) = -\Psi_{rr}(r,-\varphi), \quad \Psi_{\varphi\varphi}(r,\varphi) = -\Psi_{\varphi\varphi}(r,-\varphi), \tag{61}$$

$$\Psi_{r\varphi}(r,\varphi) = \Psi_{r\varphi}(r,-\varphi), \tag{62}$$

$$\mu_{rrr}(r,\varphi) = -\mu_{rrr}(r,-\varphi), \quad \mu_{\varphi rr}(r,\varphi) = \mu_{\varphi rr}(r,-\varphi), \tag{63}$$

$$\mu_{r\varphi\varphi}(r,\varphi) = -\mu_{r\varphi\varphi}(r,-\varphi), \quad \mu_{\varphi\varphi\varphi}(r,\varphi) = \mu_{\varphi\varphi\varphi}(r,-\varphi), \tag{64}$$

$$\mu_{rzz}(r,\varphi) = -\mu_{rzz}(r,-\varphi), \quad \mu_{\varphi zz}(r,\varphi) = \mu_{\varphi zz}(r,-\varphi), \tag{65}$$

$$\mu_{rr\varphi}(r,\varphi) = -\mu_{rr\varphi}(r,-\varphi), \quad \mu_{\varphi r\varphi}(r,\varphi) = \mu_{\varphi r\varphi}(r,-\varphi). \tag{66}$$

3. Near-tip asymptotic solutions for Mode-I and Mode-II loading conditions

We shall solve the given problems by employing asymptotic expansions of Williams' type (see Williams [3]).

3.1 Expansions of Williams' type

As the components of the macrodisplacement and the microdeformation reflect the independent kinematical degrees of freedom, we assume for u_α and $\Psi_{\alpha\beta}$ asymptotic expansions of the same form. We fix the crack tip at the origin O of the coordinate system (see Figure 1 in Part I) and set

$$u_\alpha = r^p\, u_\alpha^{(0)} + r^{p+\frac{1}{2}} u_\alpha^{(1)} + \ldots = \sum_{k=0}^{\infty} r^{p+\frac{k}{2}} u_\alpha^{(k)}, \tag{67}$$

$$\Psi_{\alpha\beta} = \overline{\Psi}_{\alpha\beta} + r^p\, \Psi_{\alpha\beta}^{(0)} + r^{p+\frac{1}{2}} \Psi_{\alpha\beta}^{(1)} + \ldots = \overline{\Psi}_{\alpha\beta} + \sum_{k=0}^{\infty} r^{p+\frac{k}{2}} \Psi_{\alpha\beta}^{(k)}, \tag{68}$$

with

$$u_\alpha^{(k)} = u_\alpha^{(k)}(\varphi), \quad \Psi_{\alpha\beta}^{(k)} = \Psi_{\alpha\beta}^{(k)}(\varphi), \quad \overline{\Psi}_{\alpha\beta} = \overline{\Psi}_{\alpha\beta}(\varphi), \tag{69}$$

and p being a real number. Since the crack tip is fixed at O, no constant term is present in the expansion of $\boldsymbol{u}$ in Eq. (67). However, we allow a constant term $\overline{\boldsymbol{\Psi}} =$ const., with physical components $\overline{\Psi}_{\alpha\beta}$ in conjunction with cylindrical coordinates, to be present in the expansion of $\boldsymbol{\Psi}$. While the Cartesian components $\overline{\Psi}_{ij}$ are constant, the physical components $\overline{\Psi}_{\alpha\beta}$ are functions of φ. There are the following well known transformation rules between $\overline{\Psi}_{\alpha\beta}$ and $\overline{\Psi}_{ij}$ (see any textbook)

$$\overline{\Psi}_{rr} = \frac{1}{2}\left(\overline{\Psi}_{11} + \overline{\Psi}_{22}\right) + \frac{1}{2}\left(\overline{\Psi}_{11} - \overline{\Psi}_{22}\right)\cos 2\varphi + \overline{\Psi}_{12}\sin 2\varphi, \tag{70}$$

$$\overline{\Psi}_{\varphi\varphi} = \frac{1}{2}\left(\overline{\Psi}_{11} + \overline{\Psi}_{22}\right) - \frac{1}{2}\left(\overline{\Psi}_{11} - \overline{\Psi}_{22}\right)\cos 2\varphi - \overline{\Psi}_{12}\sin 2\varphi, \tag{71}$$

$$\overline{\Psi}_{r\varphi} = -\frac{1}{2}\left(\overline{\Psi}_{11} - \overline{\Psi}_{22}\right)\sin 2\varphi + \overline{\Psi}_{12}\cos 2\varphi. \tag{72}$$

For later reference, we note the relations

$$\partial_\varphi \overline{\Psi}_{rr} - 2\overline{\Psi}_{r\varphi} = 0, \quad \partial_\varphi \overline{\Psi}_{\varphi\varphi} + 2\overline{\Psi}_{r\varphi} = 0, \tag{73}$$

$$\partial_\varphi \overline{\Psi}_{r\varphi} + \overline{\Psi}_{rr} - \overline{\Psi}_{\varphi\varphi} = 0, \tag{74}$$

which imply that the physical components $\overline{\Psi}_{\alpha\beta}$ trivially obey the nonclassical boundary conditions (32)–(35). Anticipating the results in Section 5, we decompose $\overline{\boldsymbol{\Psi}}$ into parts $\overline{\boldsymbol{\Psi}}^{I}$ and $\overline{\boldsymbol{\Psi}}^{II}$, reflecting symmetries according to Mode-I and Mode-II:

$$\overline{\Psi}_{\alpha\beta} = \overline{\Psi}^{I}_{\alpha\beta} + \overline{\Psi}^{II}_{\alpha\beta}, \tag{75}$$

with

$$\overline{\Psi}^{I}_{rr} := \overline{L}_{I,1} + \overline{L}_{I,2}\cos 2\varphi, \quad \overline{\Psi}^{II}_{rr} := \overline{L}_{II}\sin 2\varphi, \tag{76}$$

$$\overline{\Psi}^{I}_{\varphi\varphi} := \overline{L}_{I,1} - \overline{L}_{I,2}\cos 2\varphi, \quad \overline{\Psi}^{II}_{\varphi\varphi} := -\overline{L}_{II}\sin 2\varphi, \tag{77}$$

$$\overline{\Psi}^{I}_{r\varphi} := -\overline{L}_{I,2}\sin 2\varphi, \quad \overline{\Psi}^{II}_{r\varphi} := \overline{L}_{II}\cos 2\varphi \tag{78}$$

and

$$\overline{L}_{I,1} := \frac{1}{2}\left(\overline{\Psi}_{11} + \overline{\Psi}_{22}\right), \quad \overline{L}_{I,2} := \frac{1}{2}\left(\overline{\Psi}_{11} - \overline{\Psi}_{22}\right), \quad \overline{L}_{II} := \overline{\Psi}_{12}. \tag{79}$$

The main idea in Williams' approach is to expand each field variable $f(r,\varphi)$ in a sum of products as in Eqs. (67) and (68). We say that f is of the order r^m, and write $f \sim r^m$, whenever r^m is the power function of r in the leading term of the expansion of f. It can be deduced, from Eq. (67), that $\varepsilon_{\alpha\beta} \sim r^{p-1}$. From this, in turn, together with Eq. (68) and the elasticity laws (3)–(5), we can deduce, that $\Sigma_{\alpha\beta} \sim r^{p-1}$. Thus,

$$\Sigma_{\alpha\beta} = r^{p-1}\Sigma^{(0)}_{\alpha\beta} + r^{p-\frac{1}{2}}\Sigma^{(1)}_{\alpha\beta} + \ldots = \sum_{k=0}^{\infty} r^{p-1+\frac{k}{2}}\Sigma^{(k)}_{\alpha\beta}, \tag{80}$$

with

$$\Sigma^{(k)}_{\alpha\beta} = \Sigma^{(k)}_{\alpha\beta}(\varphi). \tag{81}$$

Expansion (67) suggests that the necessary and sufficient condition for u_α to vanish at the crack tip is

$$p > 0. \tag{82}$$

This restriction is in agreement with energetic considerations. To verify, we remark that $\nabla\overline{\mathbf{\Psi}} = \mathbf{0}$, as $\overline{\mathbf{\Psi}}$ is constant. Therefore, from Eq. (68) together with Eqs. (25)–(29), we may infer that $(\nabla\mathbf{\Psi})_{\alpha\beta\gamma} \sim r^{p-1}$. For the free energy per unit macrovolume ψ it follows that $\psi \sim r^{2p-2}$ [cf. Eq. (1)]. Now, consider a small circular area $r \leq R$, enclosing the crack tip. The total free energy (per unit length in z–direction) of this area is

$$\int_0^{2\pi}\int_0^R \psi\, r\, \mathrm{d}r\, \mathrm{d}\varphi. \tag{83}$$

Since $\psi r \sim r^{2p-1}$, restriction (82) is the necessary and sufficient condition for the energy in Eq. (83) to be bounded.

3.2 Consequences of the classical equilibrium equations

Substitute the expansion (80) into Eqs. (30) and (31) and collect coefficients of like powers of r, to obtain

$$\begin{aligned}
&r^{p-2}\left\{p\,\Sigma^{(0)}_{rr} + \partial_\varphi\Sigma^{(0)}_{r\varphi} - \Sigma^{(0)}_{\varphi\varphi}\right\}\\
&+r^{p-\frac{3}{2}}\left\{\left(p+\frac{1}{2}\right)\Sigma^{(1)}_{rr} + \partial_\varphi\Sigma^{(1)}_{r\varphi} - \Sigma^{(1)}_{\varphi\varphi}\right\}\\
&+r^{p-1}\left\{(p+1)\,\Sigma^{(2)}_{rr} + \partial_\varphi\,\Sigma^{(2)}_{r\varphi} - \Sigma^{(0)}_{\varphi\varphi}\right\}\\
&+\ldots = 0.
\end{aligned} \tag{84}$$

Similarly, we find from Eq. (31) that

$$\begin{aligned}
&r^{p-2}\left\{(p+1)\,\Sigma^{(0)}_{r\varphi} + \partial_\varphi\Sigma^{(0)}_{\varphi\varphi}\right\}\\
&+r^{p-\frac{3}{2}}\left\{\left(p+\frac{3}{2}\right)\Sigma^{(1)}_{r\varphi} + \partial_\varphi\Sigma^{(1)}_{\varphi\varphi}\right\}\\
&+r^{p-1}\left\{(p+2)\,\Sigma^{(2)}_{r\varphi} + \partial_\varphi\Sigma^{(2)}_{\varphi\varphi}\right\}\\
&+\ldots = 0.
\end{aligned} \tag{85}$$

3.3 Consequences of the classical compatibility condition

A look at $\chi_1(\cdot)$ in Eq. (40) reveals that χ_1 is a linear differential operator, i. e.,

$$\chi_1(\Psi_{\alpha\beta} - \overline{\Psi}_{\alpha\beta}) = \chi_1(\Psi_{\alpha\beta}) - \chi_1(\overline{\Psi}_{\alpha\beta}). \tag{86}$$

Since $\overline{\Psi}_{\alpha\beta}$ is independent of r, we infer from Eq. (40) that

$$\chi_1(\overline{\Psi}_{\alpha\beta}) = \frac{c_2 - c_1}{c_2}\frac{1}{r^2}\,\partial_\varphi(\partial_\varphi\overline{\Psi}_{rr} - 2\overline{\Psi}_{r\varphi}), \tag{87}$$

and by virtue of Eq. (73),

$$\chi_1(\overline{\Psi}_{\alpha\beta}) = 0. \tag{88}$$

Therefore, from Eq. (86),

$$\chi_1(\Psi_{\alpha\beta}) = \chi_1(\Psi_{\alpha\beta} - \overline{\Psi}_{\alpha\beta}), \tag{89}$$

and by appealing to expansion (68), we infer from Eq. (40) that

$$\begin{aligned}\chi_1(\Psi_{\alpha\beta}) = {} & \frac{c_2 - c_1}{c_2} \sum_{k=0}^{\infty} r^{p-2+\frac{k}{2}} \left\{ \left(p + \frac{k}{2}\right)\left(p + \frac{k}{2} - 1\right) \Psi_{\varphi\varphi}^{(k)} + \partial_{\varphi\varphi} \Psi_{rr}^{(k)} \right. \\ & -2\left(p + \frac{k}{2}\right) \partial_\varphi \Psi_{r\varphi}^{(k)} - \left(p + \frac{k}{2}\right) \Psi_{rr}^{(k)} \\ & \left. +2\left(p + \frac{k}{2}\right) \Psi_{\varphi\varphi}^{(k)} - 2\partial_\varphi \Psi_{r\varphi}^{(k)} \right\}.\end{aligned} \tag{90}$$

Similarly, by appealing to expansion (80), we infer from Eq. (41) that

$$\begin{aligned}\chi_2(\Sigma_{\alpha\beta}) = {} & \frac{(1-\nu)c_1}{2\mu c_2} \sum_{k=0}^{\infty} r^{p-2+\frac{k}{2}} \left\{ \left(p - 1 + \frac{k}{2}\right)\left(p - 2 + \frac{k}{2}\right)\left(\Sigma_{rr}^{(k)} + \Sigma_{\varphi\varphi}^{(k)}\right) \right. \\ & +\partial_{\varphi\varphi}\left(\Sigma_{rr}^{(k)} + \Sigma_{\varphi\varphi}^{(k)}\right) \\ & \left. +\left(p - 1 + \frac{k}{2}\right)\left(\Sigma_{rr}^{(k)} + \Sigma_{\varphi\varphi}^{(k)}\right) \right\}.\end{aligned} \tag{91}$$

Inserting Eqs. (90) and (91) into Eq. (39) and collecting coefficients of like powers of r gives, after some lengthy but straightforward manipulations,

$$\begin{aligned}& r^{p-3} \frac{(1-\nu)c_1}{2\mu c_2} \left\{ (p-1)^2 \left(\Sigma_{rr}^{(0)} + \Sigma_{\varphi\varphi}^{(0)}\right) + \partial_{\varphi\varphi}\left(\Sigma_{rr}^{(0)} + \Sigma_{\varphi\varphi}^{(0)}\right) \right\} \\ & +r^{p-\frac{5}{2}} \frac{(1-\nu)c_1}{2\mu c_2} \left\{ \left(p - \frac{1}{2}\right)^2 \left(\Sigma_{rr}^{(1)} + \Sigma_{\varphi\varphi}^{(1)}\right) + \partial_{\varphi\varphi}\left(\Sigma_{rr}^{(1)} + \Sigma_{\varphi\varphi}^{(1)}\right) \right\} \\ & +r^{p-2} \left\{ \frac{(1-\nu)c_1}{2\mu c_2} \left[p^2 \left(\Sigma_{rr}^{(2)} + \Sigma_{\varphi\varphi}^{(2)}\right) + \partial_{\varphi\varphi}\left(\Sigma_{rr}^{(2)} + \Sigma_{\varphi\varphi}^{(2)}\right) \right] \right. \\ & \quad \left. + \frac{c_2 - c_1}{c_2} \left[p(p+1) \Psi_{\varphi\varphi}^{(0)} + \partial_{\varphi\varphi} \Psi_{rr}^{(0)} - 2(p+1) \Psi_{r\varphi}^{(0)} \right] \right\} \\ & +r^{p-\frac{3}{2}} \left\{ \frac{(1-\nu)c_1}{2\mu c_2} \left[\left(p + \frac{1}{2}\right)^2 \left(\Sigma_{rr}^{(3)} + \Sigma_{\varphi\varphi}^{(3)}\right) + \partial_{\varphi\varphi}\left(\Sigma_{rr}^{(3)} + \Sigma_{\varphi\varphi}^{(3)}\right) \right] \right. \\ & \quad \left. + \frac{c_2 - c_1}{c_2} \left[\left(p + \frac{1}{2}\right)\left(p + \frac{3}{2}\right) \Psi_{\varphi\varphi}^{(1)} + \partial_{\varphi\varphi} \Psi_{rr}^{(1)} - 2\left(p + \frac{3}{2}\right) \Psi_{r\varphi}^{(1)} \right] \right\} \\ & + \ldots = 0.\end{aligned} \tag{92}$$

3.4 Consequences of the classical boundary conditions

By invoking the asymptotic expansion (80) in the classical boundary conditions (45) and (46), we conclude that

$$r^{p-1} \left[\Sigma_{r\varphi}^{(0)}\right]_{\varphi=\pm\pi} + r^{p-\frac{1}{2}} \left[\Sigma_{r\varphi}^{(1)}\right]_{\varphi=\pm\pi} + \ldots = 0, \tag{93}$$

$$r^{p-1} \left[\Sigma_{\varphi\varphi}^{(0)}\right]_{\varphi=\pm\pi} + r^{p-\frac{1}{2}} \left[\Sigma_{\varphi\varphi}^{(1)}\right]_{\varphi=\pm\pi} + \ldots = 0. \tag{94}$$

3.5 Cauchy stress

Before going any further, it is convenient to evaluate the results so far. The necessary and sufficient conditions for the equilibrium Eqs. (84) and (85), the compatibility condition (92) and the boundary conditions (93) and (94) to hold for arbitrary r in the vicinity of the crack tip are vanishing coefficients of all powers of r. For $\Sigma_{\alpha\beta}^{(k)}, k = 0, 1, 2$, this leads to the following systems of differential equations and associated boundary conditions.

Terms $\Sigma_{\alpha\beta}^{(0)}$

$$\partial_\varphi \Sigma_{r\varphi}^{(0)} + p\,\Sigma_{rr}^{(0)} - \Sigma_{\varphi\varphi}^{(0)} = 0, \tag{95}$$

$$\partial_\varphi \Sigma_{\varphi\varphi}^{(0)} + (p+1)\,\Sigma_{r\varphi}^{(0)} = 0, \tag{96}$$

$$\partial_{\varphi\varphi}\left(\Sigma_{rr}^{(0)} + \Sigma_{\varphi\varphi}^{(0)}\right) + (p-1)^2\left(\Sigma_{rr}^{(0)} + \Sigma_{\varphi\varphi}^{(0)}\right) = 0, \tag{97}$$

with boundary conditions

$$\left[\Sigma_{r\varphi}^{(0)}\right]_{\varphi=\pm\pi} = 0, \quad \left[\Sigma_{\varphi\varphi}^{(0)}\right]_{\varphi=\pm\pi} = 0. \tag{98}$$

Terms $\Sigma_{\alpha\beta}^{(1)}$

$$\partial_\varphi \Sigma_{r\varphi}^{(1)} + \left(p + \frac{1}{2}\right)\Sigma_{rr}^{(1)} - \Sigma_{\varphi\varphi}^{(1)} = 0, \tag{99}$$

$$\partial_\varphi \Sigma_{\varphi\varphi}^{(1)} + \left(p + \frac{3}{2}\right)\Sigma_{r\varphi}^{(1)} = 0, \tag{100}$$

$$\partial_{\varphi\varphi}\left(\Sigma_{rr}^{(1)} + \Sigma_{\varphi\varphi}^{(1)}\right) + \left(p - \frac{1}{2}\right)^2\left(\Sigma_{rr}^{(1)} + \Sigma_{\varphi\varphi}^{(1)}\right) = 0, \tag{101}$$

with boundary conditions

$$\left[\Sigma_{r\varphi}^{(1)}\right]_{\varphi=\pm\pi} = 0, \quad \left[\Sigma_{\varphi\varphi}^{(1)}\right]_{\varphi=\pm\pi} = 0. \tag{102}$$

Terms $\Sigma_{\alpha\beta}^{(2)}$

$$\partial_\varphi \Sigma_{r\varphi}^{(2)} + (p+1)\,\Sigma_{rr}^{(2)} - \Sigma_{\varphi\varphi}^{(2)} = 0, \tag{103}$$

$$\partial_\varphi \Sigma_{\varphi\varphi}^{(2)} + (p+2)\,\Sigma_{r\varphi}^{(2)} = 0, \tag{104}$$

$$\begin{aligned} &\frac{1-\nu}{2\mu c_2}\left\{\partial_{\varphi\varphi}\left(\Sigma_{rr}^{(2)} + \Sigma_{\varphi\varphi}^{(2)}\right) + p^2\left(\Sigma_{rr}^{(2)} + \Sigma_{\varphi\varphi}^{(2)}\right)\right\} \\ &+ \frac{c_2 - c_1}{c_2}\left\{\partial_{\varphi\varphi}\Psi_{\varphi\varphi}^{(0)} - 2(p+1)\,\partial_\varphi \Psi_{r\varphi}^{(0)} + p\,(p+1)\,\Psi_{\varphi\varphi}^{(0)} - p\,\Psi_{rr}^{(0)}\right\} = 0, \end{aligned} \tag{105}$$

with boundary conditions

$$\left[\Sigma_{r\varphi}^{(2)}\right]_{\varphi=\pm\pi} = 0, \quad \left[\Sigma_{\varphi\varphi}^{(2)}\right]_{\varphi=\pm\pi} = 0. \tag{106}$$

It can be recognized that coupling between components of $\boldsymbol{\Sigma}$ and components of $\boldsymbol{\Psi}$ arises for the first time in the equations for $\Sigma_{\alpha\beta}^{(2)}$. Therefore, we shall focus attention only on the terms $\Sigma_{\alpha\beta}^{(0)}$ and $\Sigma_{\alpha\beta}^{(1)}$.

The solution of the systems of differential equations for $\Sigma_{\alpha\beta}^{(0)}$ and $\Sigma_{\alpha\beta}^{(1)}$, subjected to the restriction (82), can be established by well known methods (see, e. g., A) and turns out to be identical to the solution of the corresponding problems in classical elasticity. That means that the stress components $\Sigma_{\alpha\beta}$ are singular, with order of singularity $r^{-\frac{1}{2}}$, or equivalently,

$$p = \frac{1}{2}. \tag{107}$$

The coefficients of the singular terms, $\Sigma_{\alpha\beta}^{(0)}$, are given by

$$\Sigma_{\alpha\beta}^{(0)} = \frac{\tilde{K}_I}{\sqrt{2\pi}} f^I_{\alpha\beta}(\varphi) + \frac{\tilde{K}_{II}}{\sqrt{2\pi}} f^{II}_{\alpha\beta}(\varphi), \tag{108}$$

where the constants $\tilde{K}_I$ and $\tilde{K}_{II}$ are the stress intensity factors. Here and in the following, the indices I and II stand for Mode-I and Mode-II, respectively. Moreover, we use the notations $\tilde{K}_I$ and $\tilde{K}_{II}$, in order to distinguish the stress intensity factors of the microstrain continuum from the stress intensity factors K_I and K_{II} of classical continua.

The so-called angular functions $f^I_{\alpha\beta}$ and $f^{II}_{\alpha\beta}$ are defined through

$$\begin{pmatrix} f^I_{rr} \\ f^I_{\varphi\varphi} \\ f^I_{r\varphi} \end{pmatrix} = \frac{1}{4} \begin{pmatrix} 5\cos\frac{\varphi}{2} - \cos\frac{3\varphi}{2} \\ 3\cos\frac{\varphi}{2} + \cos\frac{3\varphi}{2} \\ \sin\frac{\varphi}{2} + \sin\frac{3\varphi}{2} \end{pmatrix}, \quad \begin{pmatrix} f^{II}_{rr} \\ f^{II}_{\varphi\varphi} \\ f^{II}_{r\varphi} \end{pmatrix} = \frac{1}{4} \begin{pmatrix} -5\sin\frac{\varphi}{2} + 3\sin\frac{3\varphi}{2} \\ -3\sin\frac{\varphi}{2} - 3\sin\frac{3\varphi}{2} \\ \cos\frac{\varphi}{2} + 3\cos\frac{3\varphi}{2} \end{pmatrix}, \tag{109}$$

and are normalized by the conditions

$$\left[f^I_{\varphi\varphi}\right]_{\varphi=0} = 1, \quad \left[f^{II}_{r\varphi}\right]_{\varphi=0} = 1. \tag{110}$$

The constant terms $\Sigma_{\alpha\beta}^{(1)}$ are given by

$$\begin{pmatrix} \Sigma_{rr}^{(1)} \\ \Sigma_{\varphi\varphi}^{(1)} \\ \Sigma_{r\varphi}^{(1)} \end{pmatrix} = \tilde{k}_I \begin{pmatrix} \cos^2\varphi \\ \sin^2\varphi \\ -\frac{1}{2}\sin 2\varphi \end{pmatrix} \tag{111}$$

with $\tilde{k}_I$ being constant. Constant terms for Mode-II are not present. The first two terms of the asymptotic expansion of $\Sigma_{\alpha\beta}$ are summarized in Section 5.

3.6 Strain

Although the first two terms in the expansion of $\Sigma_{\alpha\beta}$ are identical to the ones of classical elasticity, the corresponding terms of $\varepsilon_{\alpha\beta}$ differ from those of classical elasticity. This follows from the fact that the elasticity laws (3)–(5) are not classical.

Evidently, the components $\varepsilon_{\alpha\beta}$ obey the asymptotic expansion

$$\varepsilon_{\alpha\beta} = r^{-\frac{1}{2}}\varepsilon_{\alpha\beta}^{(0)} + \varepsilon_{\alpha\beta}^{(1)} + \dots . \tag{112}$$

We use this and the asymptotic expansions of Section 3.1, with $p = \frac{1}{2}$, in the elasticity laws (3)–(5), and collect coefficients of like powers of r. Thus, we derive the following solutions for $\varepsilon_{\alpha\beta}^{(0)}$ and $\varepsilon_{\alpha\beta}^{(1)}$.

Terms $\varepsilon_{\alpha\beta}^{(0)}$

$$\varepsilon_{rr}^{(0)} = \frac{c_1}{2\mu c_2}\left[\Sigma_{rr}^{(0)} - \nu\left(\Sigma_{rr}^{(0)} + \Sigma_{\varphi\varphi}^{(0)}\right)\right], \tag{113}$$

$$\varepsilon_{\varphi\varphi}^{(0)} = \frac{c_1}{2\mu c_2}\left[\Sigma_{\varphi\varphi}^{(0)} - \nu\left(\Sigma_{rr}^{(0)} + \Sigma_{\varphi\varphi}^{(0)}\right)\right], \tag{114}$$

$$\varepsilon_{r\varphi}^{(0)} = \frac{c_1}{2\mu c_2}\Sigma_{r\varphi}^{(0)}. \tag{115}$$

By taking into account the solutions for $\Sigma_{\alpha\beta}^{(0)}$ of the last section, we find that

$$\begin{aligned}\varepsilon_{rr}^{(0)} &= \frac{c_1\tilde{K}_I}{8\mu c_2\sqrt{2\pi}}\left[(5-8\nu)\cos\frac{\varphi}{2} - \cos\frac{3\varphi}{2}\right]\\ &+\frac{c_1\tilde{K}_{II}}{8\mu c_2\sqrt{2\pi}}\left[-(5-8\nu)\sin\frac{\varphi}{2} + 3\sin\frac{3\varphi}{2}\right],\end{aligned} \tag{116}$$

$$\begin{aligned}\varepsilon_{\varphi\varphi}^{(0)} &= \frac{c_1\tilde{K}_I}{8\mu c_2\sqrt{2\pi}}\left[(3-8\nu)\cos\frac{\varphi}{2} + \cos\frac{3\varphi}{2}\right]\\ &+\frac{c_1\tilde{K}_{II}}{8\mu c_2\sqrt{2\pi}}\left[-(3-8\nu)\sin\frac{\varphi}{2} - 3\sin\frac{3\varphi}{2}\right],\end{aligned} \tag{117}$$

$$\varepsilon_{r\varphi}^{(0)} = \frac{c_1\tilde{K}_I}{8\mu c_2\sqrt{2\pi}}\left[\sin\frac{\varphi}{2} + \sin\frac{3\varphi}{2}\right] + \frac{c_1\tilde{K}_{II}}{8\mu c_2\sqrt{2\pi}}\left[\cos\frac{\varphi}{2} + 3\cos\frac{3\varphi}{2}\right]. \tag{118}$$

Terms $\varepsilon_{\alpha\beta}^{(1)}$

$$\varepsilon_{rr}^{(1)} = \frac{c_1}{2\mu c_2}\left[\Sigma_{rr}^{(1)} - \nu\left(\Sigma_{rr}^{(1)} + \Sigma_{\varphi\varphi}^{(1)}\right)\right] + \frac{c_2 - c_1}{c_2}\overline{\Psi}_{rr}, \tag{119}$$

$$\varepsilon_{\varphi\varphi}^{(1)} = \frac{c_1}{2\mu c_2}\left[\Sigma_{\varphi\varphi}^{(1)} - \nu\left(\Sigma_{rr}^{(1)} + \Sigma_{\varphi\varphi}^{(1)}\right)\right] + \frac{c_2 - c_1}{c_2}\overline{\Psi}_{\varphi\varphi}, \tag{120}$$

$$\varepsilon_{r\varphi}^{(1)} = \frac{c_1}{2\mu c_2}\Sigma_{r\varphi}^{(1)} + \frac{c_2 - c_1}{c_2}\overline{\Psi}_{r\varphi}. \tag{121}$$

Now, we take into account the solutions for $\Sigma_{\alpha\beta}^{(1)}$, established in the last section, as well as the representations for $\overline{\Psi}_{\alpha\beta}$, given by Eqs. (75)–(79), to obtain

$$\varepsilon_{rr}^{(1)} = \tilde{k}_{I,1}^{\varepsilon} + \tilde{k}_{I,2}^{\varepsilon}\cos 2\varphi + \tilde{k}_{II}^{\varepsilon}\sin 2\varphi, \tag{122}$$

$$\varepsilon_{\varphi\varphi}^{(1)} = \tilde{k}_{I,1}^{\varepsilon} - \tilde{k}_{I,2}^{\varepsilon}\cos 2\varphi - \tilde{k}_{II}^{\varepsilon}\sin 2\varphi, \tag{123}$$

$$\varepsilon_{r\varphi}^{(1)} = -\tilde{k}_{I,2}^{\varepsilon}\sin 2\varphi + \tilde{k}_{II}^{\varepsilon}\cos 2\varphi. \tag{124}$$

The constants $\tilde{k}^{\varepsilon}_{I,1}$, $\tilde{k}^{\varepsilon}_{I,2}$ and $\tilde{k}^{\varepsilon}_{II}$ are defined as follows:

$$\tilde{k}^{\varepsilon}_{I,1} := \frac{c_1 \tilde{k}_I (1-2\nu)}{4\mu c_2} + \frac{(c_2 - c_1)\overline{L}_{I,1}}{c_2}, \tag{125}$$

$$\tilde{k}^{\varepsilon}_{I,2} := \frac{c_1 \tilde{k}_I}{4\mu c_2} + \frac{(c_2 - c_1)\overline{L}_{I,2}}{c_2}, \tag{126}$$

$$\tilde{k}^{\vec{\varepsilon}}_{II} := \frac{(c_2 - c_1)\overline{L}_{II}}{c_2}. \tag{127}$$

The first two terms of the asymptotic expansion of $\varepsilon_{\alpha\beta}$ are summarized in Section 5.

3.7 Macrodisplacements

The macrodisplacement components u_r and u_φ will be determined by integrating Eqs. (24). For plane strain elasticity, it is well known that the constants of integration represent rigid body motions. Omitting such motions, we conclude for the radial component u_r that

$$u_r = \int \varepsilon_{rr} \mathrm{d}r = \int \left(r^{-\frac{1}{2}} \varepsilon^{(0)}_{rr} + \varepsilon^{(1)}_{rr} + \ldots \right) \mathrm{d}r, \tag{128}$$

or

$$r^{\frac{1}{2}} u^{(0)}_r + r u^{(1)}_r + \ldots = 2 r^{\frac{1}{2}} \varepsilon^{(0)}_{rr} + r \varepsilon^{(1)}_{rr} + \ldots . \tag{129}$$

For the circumferential component u_φ, we conclude that

$$u_\varphi = \int (r \varepsilon_{\varphi\varphi} - u_r) \, \mathrm{d}\varphi, \tag{130}$$

or

$$r^{\frac{1}{2}} u^{(0)}_\varphi + r u^{(1)}_\varphi + \ldots = r^{\frac{1}{2}} \int \left(\varepsilon^{(0)}_{\varphi\varphi} - u^{(0)}_r \right) \mathrm{d}\varphi + r \int \left(\varepsilon^{(1)}_{\varphi\varphi} - u^{(1)}_r \right) \mathrm{d}\varphi + \ldots . \tag{131}$$

By employing steps similar to those in the last section, we get the following solutions for $u^{(0)}_\alpha$ and $u^{(1)}_\alpha$.

Terms $u^{(0)}_\alpha$

$$u^{(0)}_r = 2 \varepsilon^{(0)}_{rr}, \tag{132}$$

$$u^{(0)}_\varphi = \int \left(\varepsilon^{(0)}_{\varphi\varphi} - u^{(0)}_r \right) \mathrm{d}\varphi = \int \left(\varepsilon^{(0)}_{\varphi\varphi} - 2 \varepsilon^{(0)}_{rr} \right) \mathrm{d}\varphi. \tag{133}$$

Invoking Eqs. (116) and (117), we get, after some straightforward manipulations,

$$u^{(0)}_r = \frac{c_1 \tilde{K}_I}{4\mu c_2 \sqrt{2\pi}} \left[(5 - 8\nu) \cos \frac{\varphi}{2} - \cos \frac{3\varphi}{2} \right]$$

$$+\frac{c_1\tilde{K}_{II}}{4\mu c_2\sqrt{2\pi}}\left[-(5-8\nu)\sin\frac{\varphi}{2}+3\sin\frac{3\varphi}{2}\right], \tag{134}$$

$$u_\varphi^{(0)}=\frac{c_1\tilde{K}_I}{4\mu c_2\sqrt{2\pi}}\left[-(7-8\nu)\sin\frac{\varphi}{2}+\sin\frac{3\varphi}{2}\right]$$
$$+\frac{c_1\tilde{K}_{II}}{4\mu c_2\sqrt{2\pi}}\left[-(7-8\nu)\cos\frac{\varphi}{2}+3\cos\frac{3\varphi}{2}\right]. \tag{135}$$

Terms $u_\alpha^{(1)}$

$$u_r^{(1)}=\varepsilon_{rr}^{(1)}, \tag{136}$$

$$u_\varphi^{(1)}=\int\left(\varepsilon_{\varphi\varphi}^{(1)}-u_r^{(1)}\right)\mathrm{d}\varphi=\int\left(\varepsilon_{\varphi\varphi}^{(1)}-\varepsilon_{rr}^{(1)}\right)\mathrm{d}\varphi, \tag{137}$$

from which, by virtue of Eqs. (119) and (120),

$$u_r^{(1)}=\tilde{k}_{I,1}^{\vec{\varepsilon}}+\tilde{k}_{I,2}^{\vec{\varepsilon}}\cos 2\varphi+\tilde{k}_{II}^{\vec{\varepsilon}}\sin 2\varphi, \tag{138}$$

$$u_\varphi^{(1)}=-\tilde{k}_{I,2}^{\vec{\varepsilon}}\sin 2\varphi+\tilde{k}_{II}^{\vec{\varepsilon}}\cos 2\varphi. \tag{139}$$

The first two terms of the asymptotic expansion of u_α are also summarized in Section 5.

3.8 Microdeformation

We shall derive differential equations for $\Psi_{\alpha\beta}^{(0)}$ and $\Psi_{\alpha\beta}^{(1)}$ by inserting the asymptotic expansions of $\Psi_{\alpha\beta}$ and $\Sigma_{\alpha\beta}$ (see Eqs. (68) and (80), with $p=\frac{1}{2}$) into Eqs. (36)–(38). Note that, by virtue of Eqs. (73) and (74) and since $\overline{\Psi}_{\alpha\beta}$ is independent of r, the identity

$$\partial_{rr}\overline{\Psi}_{rr}+\frac{1}{r^2}\partial_{\varphi\varphi}\overline{\Psi}_{rr}+\frac{1}{r}\partial_r\overline{\Psi}_{rr}-\frac{4}{r^2}\partial_\varphi\overline{\Psi}_{r\varphi}-\frac{2}{r^2}\overline{\Psi}_{rr}+\frac{2}{r^2}\overline{\Psi}_{\varphi\varphi}=0 \tag{140}$$

applies. Keeping this in mind and collecting terms of like powers of r, after some lengthy but otherwise straightforward manipulations, Eq. (36) yields

$$\begin{aligned}&r^{-\frac{3}{2}}\left\{\partial_{\varphi\varphi}\Psi_{rr}^{(0)}-4\partial_\varphi\Psi_{r\varphi}^{(0)}-\frac{7}{4}\Psi_{rr}^{(0)}+2\Psi_{\varphi\varphi}^{(0)}\right\}\\&+r^{-1}\left\{\partial_{\varphi\varphi}\Psi_{rr}^{(1)}-4\partial_\varphi\Psi_{r\varphi}^{(1)}-\Psi_{rr}^{(1)}+2\Psi_{\varphi\varphi}^{(1)}\right\}+\ldots=0.\end{aligned} \tag{141}$$

Similarly, from Eqs. (37) and (38), we get

$$\begin{aligned}&r^{-\frac{3}{2}}\left\{\partial_{\varphi\varphi}\Psi_{\varphi\varphi}^{(0)}+4\partial_\varphi\Psi_{r\varphi}^{(0)}-\frac{7}{4}\Psi_{\varphi\varphi}^{(0)}+2\Psi_{rr}^{(0)}\right\}\\&+r^{-1}\left\{\partial_{\varphi\varphi}\Psi_{\varphi\varphi}^{(1)}+4\partial_\varphi\Psi_{r\varphi}^{(1)}-\Psi_{\varphi\varphi}^{(1)}+2\Psi_{rr}^{(1)}\right\}+\ldots=0,\end{aligned} \tag{142}$$

$$\begin{aligned}&r^{-\frac{3}{2}}\left\{\partial_{\varphi\varphi}\Psi_{r\varphi}^{(0)}+2\partial_\varphi\left[\Psi_{rr}^{(0)}-\Psi_{\varphi\varphi}^{(0)}\right]-\frac{15}{4}\Psi_{r\varphi}^{(0)}\right\}\\&+r^{-1}\left\{\partial_{\varphi\varphi}\Psi_{r\varphi}^{(1)}+2\partial_\varphi\left[\Psi_{rr}^{(0)}-\Psi_{\varphi\varphi}^{(0)}\right]-\frac{15}{4}\Psi_{r\varphi}^{(0)}\right\}+\ldots=0.\end{aligned} \tag{143}$$

It is worth remarking that if only terms up to order r^{-1} are retained in Eqs. (141)–(143), then the terms $\Psi^{(0)}_{\alpha\beta}$ and $\Psi^{(1)}_{\alpha\beta}$ are uncoupled from the terms $\overline{\Psi}_{\alpha\beta}$ and $\Sigma^{(k)}_{\alpha\beta}$.

In an analogous manner, by substituting the asymptotic expansion (68) into the nonclassical boundary conditions (48)–(50), we show that

$$r^{\frac{1}{2}}\left[\partial_\varphi \Psi^{(0)}_{rr} - 2\Psi^{(0)}_{r\varphi}\right]_{\varphi=\pm\pi} + r\left[\partial_\varphi \Psi^{(1)}_{rr} - 2\Psi^{(1)}_{r\varphi}\right]_{\varphi=\pm\pi} + \ldots = 0, \tag{144}$$

$$r^{\frac{1}{2}}\left[\partial_\varphi \Psi^{(0)}_{\varphi\varphi} + 2\Psi^{(0)}_{r\varphi}\right]_{\varphi=\pm\pi} + r\left[\partial_\varphi \Psi^{(1)}_{\varphi\varphi} + 2\Psi^{(1)}_{r\varphi}\right]_{\varphi=\pm\pi} + \ldots = 0, \tag{145}$$

$$r^{\frac{1}{2}}\left[\partial_\varphi \Psi^{(0)}_{r\varphi} + \Psi^{(0)}_{rr} - \Psi^{(0)}_{\varphi\varphi}\right]_{\varphi=\pm\pi} + r\left[\partial_\varphi \Psi^{(1)}_{r\varphi} + \Psi^{(1)}_{rr} - \Psi^{(1)}_{\varphi\varphi}\right]_{\varphi=\pm\pi} + \ldots = 0. \tag{146}$$

3.8.1 Differential equations for $\Psi^{(0)}_{\alpha\beta}$

Equating to zero the coefficients of power $r^{-\frac{3}{2}}$ in Eqs. (141)–(143) leads to the system of ordinary differential equations

$$\partial_{\varphi\varphi}\Psi^{(0)}_{rr} - 4\partial_\varphi \Psi^{(0)}_{r\varphi} - \frac{7}{4}\Psi^{(0)}_{rr} + 2\Psi^{(0)}_{\varphi\varphi} = 0, \tag{147}$$

$$\partial_{\varphi\varphi}\Psi^{(0)}_{\varphi\varphi} + 4\partial_\varphi \Psi^{(0)}_{r\varphi} - \frac{7}{4}\Psi^{(0)}_{\varphi\varphi} + 2\Psi^{(0)}_{rr} = 0, \tag{148}$$

$$\partial_{\varphi\varphi}\Psi^{(0)}_{r\varphi} + 2\partial_\varphi\left[\Psi^{(0)}_{rr} - \Psi^{(0)}_{\varphi\varphi}\right] - \frac{15}{4}\Psi^{(0)}_{r\varphi} = 0. \tag{149}$$

Similarly, equating to zero the coefficients of power $r^{\frac{1}{2}}$ in the boundary conditions (144)–(146) leads to

$$\left[\partial_\varphi \Psi^{(0)}_{rr} - 2\Psi^{(0)}_{r\varphi}\right]_{\varphi=\pm\pi} = 0, \tag{150}$$

$$\left[\partial_\varphi \Psi^{(0)}_{\varphi\varphi} + 2\Psi^{(0)}_{r\varphi}\right]_{\varphi=\pm\pi} = 0, \tag{151}$$

$$\left[\partial_\varphi \Psi^{(0)}_{r\varphi} + \Psi^{(0)}_{rr} - \Psi^{(0)}_{\varphi\varphi}\right]_{\varphi=\pm\pi} = 0. \tag{152}$$

Proceeding to solve the system (147)–(149), we note that Eqs. (147) and (148) imply the ordinary differential equation

$$\partial_{\varphi\varphi}\left[\Psi^{(0)}_{rr} + \Psi^{(0)}_{\varphi\varphi}\right] + \frac{1}{4}\left[\Psi^{(0)}_{rr} + \Psi^{(0)}_{\varphi\varphi}\right] = 0 \tag{153}$$

for the sum $\Psi^{(0)}_{rr} + \Psi^{(0)}_{\varphi\varphi}$, which has the solution

$$\Psi^{(0)}_{rr} + \Psi^{(0)}_{\varphi\varphi} = A^{(0)}\cos\frac{\varphi}{2} + B^{(0)}\sin\frac{\varphi}{2}. \tag{154}$$

For determining the constants of integration $A^{(0)}$ and $B^{(0)}$, we utilize the boundary conditions. From Eqs. (150) and (151), we derive the equation

$$\left[\partial_\varphi\left(\Psi^{(0)}_{rr} + \Psi^{(0)}_{\varphi\varphi}\right)\right]_{\varphi=\pm\pi} = 0. \tag{155}$$

By substituting the solution (154), we see that

$$A^{(0)} = 0. \tag{156}$$

To go further, we notice that Eqs. (147) and (148) imply

$$\partial_{\varphi}\Psi_{r\varphi}^{(0)} = \frac{1}{8}\left[\partial_{\varphi\varphi}\left(\Psi_{rr}^{(0)} - \Psi_{\varphi\varphi}^{(0)}\right) - \frac{15}{4}\left(\Psi_{rr}^{(0)} - \Psi_{\varphi\varphi}^{(0)}\right)\right]. \tag{157}$$

Next, we differentiate Eq. (149) with respect to φ and use Eq. (157). Rearrangement of terms leads to the ordinary differential equation

$$\begin{aligned}&\frac{1}{2}\partial_{\varphi\varphi\varphi\varphi}\left(\Psi_{rr}^{(0)} + \Psi_{\varphi\varphi}^{(0)}\right) + \frac{17}{4}\partial_{\varphi\varphi}\left(\Psi_{rr}^{(0)} + \Psi_{\varphi\varphi}^{(0)}\right) + \frac{225}{32}\left(\Psi_{rr}^{(0)} + \Psi_{\varphi\varphi}^{(0)}\right)\\ &-\partial_{\varphi\varphi\varphi\varphi}\Psi_{\varphi\varphi}^{(0)} - \frac{17}{2}\partial_{\varphi\varphi}\Psi_{\varphi\varphi}^{(0)} - \frac{225}{16}\Psi_{\varphi\varphi}^{(0)} = 0.\end{aligned} \tag{158}$$

By substituting the solutions (154) and (156), we gain an ordinary differential equation for $\Psi_{\varphi\varphi}^{(0)}$,

$$\partial_{\varphi\varphi\varphi\varphi}\Psi_{\varphi\varphi}^{(0)} + \frac{17}{2}\partial_{\varphi\varphi}\Psi_{\varphi\varphi}^{(0)} + \frac{225}{16}\Psi_{\varphi\varphi}^{(0)} = 6B^{(0)}\sin\frac{\varphi}{2}, \tag{159}$$

which obeyes the solution

$$\begin{aligned}\Psi_{\varphi\varphi}^{(0)} = &\frac{1}{2}B^{(0)}\sin\frac{\varphi}{2} + E^{(0)}\sin\frac{3\varphi}{2} + F^{(0)}\sin\frac{5\varphi}{2} + C^{(0)}\cos\frac{3\varphi}{2}\\ &+D^{(0)}\cos\frac{5\varphi}{2},\end{aligned} \tag{160}$$

with $C^{(0)}, D^{(0)}, E^{(0)}$ and $F^{(0)}$ being new constants of integration. Further, from Eqs. (154), (156) and (160),

$$\begin{aligned}\Psi_{rr}^{(0)} = &\frac{1}{2}B^{(0)}\sin\frac{\varphi}{2} - E^{(0)}\sin\frac{3\varphi}{2} - F^{(0)}\sin\frac{5\varphi}{2} - C^{(0)}\cos\frac{3\varphi}{2}\\ &-D^{(0)}\cos\frac{5\varphi}{2}.\end{aligned} \tag{161}$$

Finally, using the solutions (161) and (160) in Eq. (157), we obtain the solution $\Psi_{r\varphi}^{(0)}$ in the form

$$\Psi_{r\varphi}^{(0)} = C^{(0)}\sin\frac{3\varphi}{2} + D^{(0)}\sin\frac{5\varphi}{2} - E^{(0)}\cos\frac{3\varphi}{2} - F^{(0)}\cos\frac{5\varphi}{2} + G^{(0)}, \tag{162}$$

where $G^{(0)}$ is a further constant of integration. For the constants of integration in the solutions (160)–(162) we can verify, by evaluating the boundary conditions (150)–(152) that

$$G^{(0)} = 0, \quad -D^{(0)} = C^{(0)}, \quad -F^{(0)} = E^{(0)}. \tag{163}$$

In accordance with the symmetry conditions (53) and (54) for Mode-I as well as (61) and (62) for Mode-II, we set

$$C^{(0)} \equiv C_I^{(0)}, \quad B^{(0)} \equiv B_{II}^{(0)}, \quad E^{(0)} \equiv E_{II}^{(0)}. \tag{164}$$

Then, the solutions (160)–(162) become

$$\Psi_{rr}^{(0)} = -C_I^{(0)}\left(\cos\frac{3\varphi}{2} - \cos\frac{5\varphi}{2}\right) + \frac{1}{2}B_{II}^{(0)}\sin\frac{\varphi}{2} - E_{II}^{(0)}\left(\sin\frac{3\varphi}{2} - \sin\frac{5\varphi}{2}\right), \tag{165}$$

$$\Psi_{\varphi\varphi}^{(0)} = C_I^{(0)}\left(\cos\frac{3\varphi}{2} - \cos\frac{5\varphi}{2}\right) + \frac{1}{2}B_{II}^{(0)}\sin\frac{\varphi}{2} + E_{II}^{(0)}\left(\sin\frac{3\varphi}{2} - \sin\frac{5\varphi}{2}\right), \tag{166}$$

$$\Psi_{r\varphi}^{(0)} = C_I^{(0)}\left(\sin\frac{3\varphi}{2} - \sin\frac{5\varphi}{2}\right) - E_{II}^{(0)}\left(\cos\frac{3\varphi}{2} - \cos\frac{5\varphi}{2}\right). \tag{167}$$

It is of interest to comment the following issue. Obviously not all constants of integration may be determined, because boundary conditions are prescribed only on the crack faces. Nevertheless, it is remarkable that the solutions of Mode-I include only one unknown constant, whereas the solutions of Mode-II depend on two unknown constants. We shall come back to this specific feature in the next section as well as in Section 5, while discussing the asymptotic solutions of the double stresses.

3.8.2 Differential equations for $\Psi_{\alpha\beta}^{(1)}$

Equating to zero the coefficients of power r^{-1} in Eqs. (141)–(143) and the coefficients of power r in the boundary conditions (144)–(146) furnish the system of ordinary differential equations

$$\partial_{\varphi\varphi}\Psi_{rr}^{(1)} - 4\partial_\varphi\Psi_{r\varphi}^{(1)} - \Psi_{rr}^{(1)} + 2\Psi_{\varphi\varphi}^{(1)} = 0, \tag{168}$$

$$\partial_{\varphi\varphi}\Psi_{\varphi\varphi}^{(1)} + 4\partial_\varphi\Psi_{r\varphi}^{(1)} - \Psi_{\varphi\varphi}^{(1)} + 2\Psi_{rr}^{(1)} = 0, \tag{169}$$

$$\partial_{\varphi\varphi}\Psi_{r\varphi}^{(1)} + 2\partial_\varphi\left(\Psi_{rr}^{(1)} - \Psi_{\varphi\varphi}^{(1)}\right) - 3\Psi_{r\varphi}^{(1)} = 0, \tag{170}$$

and corresponding boundary conditions

$$\left[\partial_\varphi\Psi_{rr}^{(1)} - 2\Psi_{r\varphi}^{(1)}\right]_{\varphi=\pm\pi} = 0, \tag{171}$$

$$\left[\partial_\varphi\Psi_{\varphi\varphi}^{(1)} + 2\Psi_{r\varphi}^{(1)}\right]_{\varphi=\pm\pi} = 0, \tag{172}$$

$$\left[\partial_\varphi\Psi_{r\varphi}^{(1)} + \Psi_{rr}^{(1)} - \Psi_{\varphi\varphi}^{(1)}\right]_{\varphi=\pm\pi} = 0. \tag{173}$$

Since the steps for solving the above system of differential equations are quite similar to those in the last section, we omit the details and present only the final solutions

$$\Psi_{rr}^{(1)} = \frac{1}{2}A^{(1)}\cos\varphi - D_I^{(1)}(\cos\varphi + \cos 3\varphi) - E_{II}^{(1)}\sin\varphi - F_{II}^{(1)}\sin 3\varphi, \tag{174}$$

$$\Psi_{\varphi\varphi}^{(1)} = \frac{1}{2}A^{(1)}\cos\varphi + D_I^{(1)}(\cos\varphi + \cos 3\varphi) + E_{II}^{(1)}\sin\varphi + F_{II}^{(1)}\sin 3\varphi, \tag{175}$$

$$\Psi_{r\varphi}^{(1)} = D_I^{(1)} \left(\sin\varphi + \sin 3\varphi \right) - E_{II}^{(1)} \left(\frac{1}{2} + \cos\varphi \right) - F_{II}^{(1)} \left(\frac{1}{2} - \cos 3\varphi \right). \tag{176}$$

With regard to the symmetry conditions (53), (54), (61) and (62), the constants $A_I^{(1)}, D_I^{(1)}, E_{II}^{(1)}$ and $F_{II}^{(1)}$ are attributed to loading conditions of Mode-I and Mode-II, respectively. The solutions $\Psi_{\alpha\beta}^{(0)}$ and $\Psi_{\alpha\beta}^{(1)}$ are summarized and discussed in Section 5.

3.9 Double stress

The considerations of Section 3.1, together with $p = \frac{1}{2}$ (see Eq. (107)), and the elasticity laws for $\boldsymbol{\mu}$ [see Eqs. (7)–(16)] suggest the asymptotic expansion

$$\mu_{\alpha\beta\gamma} = r^{-\frac{1}{2}} \mu_{\alpha\beta\gamma}^{(0)} + \mu_{\alpha\beta\gamma}^{(1)} + \ldots = \sum_{k=0}^{\infty} r^{-\frac{1}{2}+\frac{k}{2}} \mu_{\alpha\beta\gamma}^{(k)}, \tag{177}$$

with

$$\mu_{\alpha\beta\gamma}^{(k)} = \mu_{\alpha\beta\gamma}^{(k)}(\varphi). \tag{178}$$

The goal is to determine $\mu_{\alpha\beta\gamma}^{(0)}$ and $\mu_{\alpha\beta\gamma}^{(1)}$ by substituting the asymptotic expansion for $\Psi_{\alpha\beta}$ into the elasticity laws (7)–(16). It is readily verified that in view of the conditions (73) and (74), the terms $\overline{\Psi}_{\alpha\beta}$ of the expansion (68) will disappear in the subsequent equations. Thus, we conclude from Eqs. (7)–(16), by equating the coefficients of power $r^{-\frac{1}{2}}$ that

$$\mu_{rrr}^{(0)} = (c_2 - c_1) \left(\frac{\lambda + 2\mu}{2} \Psi_{rr}^{(0)} + \frac{\lambda}{2} \Psi_{\varphi\varphi}^{(0)} \right), \tag{179}$$

$$\mu_{r\varphi\varphi}^{(0)} = (c_2 - c_1) \left(\frac{\lambda + 2\mu}{2} \Psi_{\varphi\varphi}^{(0)} + \frac{\lambda}{2} \Psi_{rr}^{(0)} \right), \tag{180}$$

$$\mu_{rzz}^{(0)} = (c_2 - c_1) \frac{\lambda}{2} \left(\Psi_{rr}^{(0)} + \Psi_{\varphi\varphi}^{(0)} \right), \tag{181}$$

$$\mu_{rr\varphi}^{(0)} = (c_2 - c_1) \mu \Psi_{r\varphi}^{(0)}, \tag{182}$$

$$\mu_{\varphi rr}^{(0)} = (c_2 - c_1) \left[2\mu \left(\partial_\varphi \Psi_{rr}^{(0)} - 2\Psi_{r\varphi}^{(0)} \right) + \lambda \partial_\varphi \left(\Psi_{rr}^{(0)} + \Psi_{\varphi\varphi}^{(0)} \right) \right], \tag{183}$$

$$\mu_{\varphi\varphi\varphi}^{(0)} = (c_2 - c_1) \left[2\mu \left(\partial_\varphi \Psi_{\varphi\varphi}^{(0)} + 2\Psi_{r\varphi}^{(0)} \right) + \lambda \partial_\varphi \left(\Psi_{rr}^{(0)} + \Psi_{\varphi\varphi}^{(0)} \right) \right], \tag{184}$$

$$\mu_{\varphi zz}^{(0)} = (c_2 - c_1) \lambda \partial_\varphi \left(\Psi_{rr}^{(0)} + \Psi_{\varphi\varphi}^{(0)} \right), \tag{185}$$

$$\mu_{\varphi r\varphi}^{(0)} = (c_2 - c_1) 2\mu \left(\partial_\varphi \Psi_{r\varphi}^{(0)} + \Psi_{rr}^{(0)} - \Psi_{\varphi\varphi}^{(0)} \right), \tag{186}$$

and by equating the coefficients of power r^0 that

$$\mu_{rrr}^{(1)} = (c_2 - c_1) \left((\lambda + 2\mu) \Psi_{rr}^{(1)} + \lambda \Psi_{\varphi\varphi}^{(1)} \right), \tag{187}$$

$$\mu_{r\varphi\varphi}^{(1)} = (c_2 - c_1) \left((\lambda + 2\mu) \Psi_{\varphi\varphi}^{(1)} + \lambda \Psi_{rr}^{(1)} \right), \tag{188}$$

$$\mu_{rzz}^{(1)} = (c_2 - c_1) \lambda \left(\Psi_{rr}^{(1)} + \Psi_{\varphi\varphi}^{(1)} \right), \tag{189}$$

$$\mu^{(1)}_{r\varphi\varphi} = (c_2 - c_1) 2\mu \Psi^{(1)}_{r\varphi}, \tag{190}$$

$$\mu^{(1)}_{\varphi rr} = (c_2 - c_1) \left[2\mu \left(\partial_\varphi \Psi^{(1)}_{rr} - 2\Psi^{(1)}_{r\varphi}\right) + \lambda \partial_\varphi \left(\Psi^{(1)}_{rr} + \Psi^{(1)}_{\varphi\varphi}\right)\right], \tag{191}$$

$$\mu^{(1)}_{\varphi\varphi\varphi} = (c_2 - c_1) \left[2\mu \left(\partial_\varphi \Psi^{(1)}_{\varphi\varphi} + 2\Psi^{(1)}_{r\varphi}\right) + \lambda \partial_\varphi \left(\Psi^{(1)}_{rr} + \Psi^{(1)}_{\varphi\varphi}\right)\right], \tag{192}$$

$$\mu^{(1)}_{\varphi zz} = (c_2 - c_1) \lambda \partial_\varphi \left(\Psi^{(1)}_{rr} + \Psi^{(1)}_{\varphi\varphi}\right), \tag{193}$$

$$\mu^{(1)}_{r\varphi\varphi} = (c_2 - c_1) 2\mu \left(\partial_\varphi \Psi^{(1)}_{r\varphi} + \Psi^{(1)}_{rr} - \Psi^{(1)}_{\varphi\varphi}\right). \tag{194}$$

If we introduce the solutions (160)–(162) into Eqs. (179)–(186) and rearrange terms, then, for $\mu^{(0)}_{\alpha\beta\gamma}$, we obtain the representations

$$\begin{aligned}\mu^{(0)}_{rrr} = (c_2 - c_1)\Big[&-\mu C^{(0)}_I \left(\cos\frac{3\varphi}{2} - \cos\frac{5\varphi}{2}\right) + \frac{\lambda+\mu}{2} B^{(0)}_{II} \sin\frac{\varphi}{2} \\ &-\mu E^{(0)}_{II} \left(\sin\frac{3\varphi}{2} - \sin\frac{5\varphi}{2}\right)\Big],\end{aligned} \tag{195}$$

$$\begin{aligned}\mu^{(0)}_{r\varphi\varphi} = (c_2 - c_1)\Big[&\mu C^{(0)}_I \left(\cos\frac{3\varphi}{2} - \cos\frac{5\varphi}{2}\right) + \frac{\lambda+\mu}{2} B^{(0)}_{II} \sin\frac{\varphi}{2} \\ &+\mu E^{(0)}_{II} \left(\sin\frac{3\varphi}{2} - \sin\frac{5\varphi}{2}\right)\Big],\end{aligned} \tag{196}$$

$$\mu^{(0)}_{rzz} = (c_2 - c_1) \frac{\lambda}{2} B^{(0)}_{II} \sin\frac{\varphi}{2}, \tag{197}$$

$$\begin{aligned}\mu^{(0)}_{rr\varphi} = (c_2 - c_1)\Big[&\mu C^{(0)}_I \left(\sin\frac{3\varphi}{2} - \sin\frac{5\varphi}{2}\right) \\ &-\mu E^{(0)}_{II} \left(\cos\frac{3\varphi}{2} - \cos\frac{5\varphi}{2}\right)\Big],\end{aligned} \tag{198}$$

$$\begin{aligned}\mu^{(0)}_{\varphi rr} = (c_2 - c_1)\Big[&-\mu C^{(0)}_I \left(\sin\frac{3\varphi}{2} + \sin\frac{5\varphi}{2}\right) + \frac{\lambda+\mu}{2} B^{(0)}_{II} \cos\frac{\varphi}{2} \\ &+\mu E^{(0)}_{II} \left(\cos\frac{3\varphi}{2} + \cos\frac{5\varphi}{2}\right)\Big],\end{aligned} \tag{199}$$

$$\begin{aligned}\mu^{(0)}_{\varphi\varphi\varphi} = (c_2 - c_1)\Big[&\mu C^{(0)}_I \left(\sin\frac{3\varphi}{2} + \sin\frac{5\varphi}{2}\right) + \frac{\lambda+\mu}{2} B^{(0)}_{II} \cos\frac{\varphi}{2} \\ &-\mu E^{(0)}_{II} \left(\cos\frac{3\varphi}{2} + \cos\frac{5\varphi}{2}\right)\Big],\end{aligned} \tag{200}$$

$$\mu^{(0)}_{\varphi zz} = (c_2 - c_1) \frac{\lambda}{2} B^{(0)}_{II} \cos\frac{\varphi}{2}, \tag{201}$$

$$\mu^{(0)}_{\varphi r\varphi} = (c_2 - c_1)\left[-\mu C^{(0)}_I \left(\cos\frac{3\varphi}{2} + \cos\frac{5\varphi}{2}\right) - \mu E^{(0)}_{II} \left(\sin\frac{3\varphi}{2} + \sin\frac{5\varphi}{2}\right)\right]. \tag{202}$$

The fact that the solutions $\mu^{(0)}_{\alpha\beta\gamma}$ depend on two unknown constants in case of Mode-II is a characteristic property. As we shall see in Section 5, this feature leads to the existence of two stress intensity factors for the double stresses in case of Mode-II.

Using steps similar to those above we obtain for $\mu^{(1)}_{\alpha\beta\gamma}$ the representations

$$\begin{aligned}\mu^{(1)}_{rrr} = {} & (c_2 - c_1)\Big[(\lambda+\mu)A^{(1)}_I \cos\varphi - 2\mu D^{(1)}_I(\cos\varphi + \cos 3\varphi) \\ & -2\mu E^{(1)}_{II}\sin\varphi - 2\mu F^{(1)}_{II}\sin 3\varphi\Big],\end{aligned} \tag{203}$$

$$\begin{aligned}\mu^{(1)}_{r\varphi\varphi} = {} & (c_2 - c_1)\Big[(\lambda+\mu)A^{(1)}_I \cos\varphi + 2\mu D^{(1)}_I(\cos\varphi + \cos 3\varphi) \\ & +2\mu E^{(1)}_{II}\sin\varphi + 2\mu F^{(1)}_{II}\sin 3\varphi\Big],\end{aligned} \tag{204}$$

$$\mu^{(1)}_{\varphi zz} = (c_2 - c_1)A^{(1)}_I \cos\varphi, \tag{205}$$

$$\begin{aligned}\mu^{(1)}_{rr\varphi} = {} & (c_2 - c_1)\Big[2\mu D^{(1)}_I(\sin\varphi + \sin 3\varphi) \\ & - 2\mu E^{(1)}_{II}\left(\frac{1}{2} + \cos\varphi\right) - 2\mu F^{(1)}_{II}\left(\frac{1}{2} - \cos 3\varphi\right)\Big],\end{aligned} \tag{206}$$

$$\begin{aligned}\mu^{(1)}_{\varphi rr} = {} & (c_2 - c_1)\Big[-(\lambda+\mu)A^{(1)}_I \sin\varphi - 2\mu D^{(1)}_I(\sin\varphi - \sin 3\varphi) \\ & +2\mu E^{(1)}_{II}(1+\cos\varphi) - 2\mu F^{(1)}_{II}(1+\cos 3\varphi)\Big],\end{aligned} \tag{207}$$

$$\begin{aligned}\mu^{(1)}_{\varphi\varphi\varphi} = {} & (c_2 - c_1)\Big[-(\lambda+\mu)A^{(1)}_I \sin\varphi + 2\mu D^{(1)}_I(\sin\varphi - \sin 3\varphi) \\ & -2\mu E^{(1)}_{II}(1+\cos\varphi) + 2\mu F^{(1)}_{II}(1+\cos 3\varphi)\Big],\end{aligned} \tag{208}$$

$$\mu^{(1)}_{\varphi zz} = (c_2 - c_1)\lambda A^{(1)}_I \sin\varphi, \tag{209}$$

$$\begin{aligned}\mu^{(1)}_{\varphi r\varphi} = {} & (c_2 - c_1)\Big[-2\mu D^{(1)}_I(\cos\varphi - \cos 3\varphi) \\ & -2\mu E^{(1)}_{II}\sin\varphi + 2\mu F^{(1)}_{II}\sin 3\varphi\Big].\end{aligned} \tag{210}$$

Before going to discuss the obtained solutions, it is perhaps of interest to rederive the analytical solutions by an alternative approach, starting from asymptotic expansions of $\boldsymbol{\Sigma}$ and $\boldsymbol{\mu}$ rather than the asymtptotic expansions of $\boldsymbol{u}$ and $\boldsymbol{\Psi}$ used in this section.

4. Alternative approach for the determination of the near-tip fields

In Section 3 we determined the near-tip fields by starting from asymptotic expansions of the same form for the kinematical variables $\boldsymbol{u}$ and $\boldsymbol{\Psi}$ [see Eqs. (67) and (68)]. Alternatively, it is instructive to start from asymptotic expansions of the same type for the stresses $\boldsymbol{\Sigma}$ and $\boldsymbol{\mu}$, i. e.,

$$\Sigma_{\alpha\beta} = r^{p-1}\Sigma^{(0)}_{\alpha\beta} + r^{p-\frac{1}{2}}\Sigma^{(1)}_{\alpha\beta} + \dots, \tag{211}$$

$$\mu_{\alpha\beta\gamma} = r^{p-1}\mu^{(0)}_{\alpha\beta\gamma} + r^{p-\frac{1}{2}}\mu^{(1)}_{\alpha\beta\gamma} + \dots, \tag{212}$$

where $\Sigma^{(k)}_{\alpha\beta} = \Sigma^{(k)}_{\alpha\beta}(\varphi)$ and $\mu^{(k)}_{\alpha\beta\gamma} = \mu^{(k)}_{\alpha\beta\gamma}(\varphi)$, $k = 0, 1, 2, \dots$. Then, from the elasticity laws (17)–(22), we recognize that $(\nabla\boldsymbol{\Psi})_{\alpha\beta\gamma} \sim r^{p-1}$ and hence the components $\Psi_{\alpha\beta}$ are of form (67). It follows that all outcomes of sections 3.2–3.6 apply as well and, in particular, that $p = \frac{1}{2}$. Then, it remains to show, how to determine the terms $\mu^{(0)}_{\alpha\beta\gamma}$ and $\mu^{(1)}_{\alpha\beta\gamma}$. The corresponding terms of $\boldsymbol{\Psi}$ will then be established by integrating the elasticity laws (17)–(22). For the purposes of the present section, however, it suffices to demonstrate only how to determine the terms $\mu^{(0)}_{\alpha\beta\gamma}$. To this end, we shall involve the nonclassical equilibrium Eqs. (32)–(35), in conjunction with the elasticity law (6) for $\boldsymbol{\sigma}$, as well as the nonclassical compatibility conditions (42)–(44). It is necessary to involve the latter for we are directly seeking for solutions of $\mu_{\alpha\beta\gamma}$.

4.1 Nonclassical equilibrium equations

Since $\varepsilon_{\alpha\beta} \sim r^{-\frac{1}{2}}$ and $\Psi_{\alpha\beta} \sim r^0$, we recognize from the elasticity law (6) that $\sigma_{\alpha\beta} \sim r^{-\frac{1}{2}}$. On the other hand, by virtue of the expansion (212), $\partial_r \mu_{\alpha\beta\gamma} \sim r^{-\frac{3}{2}}$ and $\frac{1}{r}\mu_{\alpha\beta\gamma} \sim r^{-\frac{3}{2}}$. Therefore, up to terms of order r^{-1} there will be no contributions of $\boldsymbol{\sigma}$ present in the nonclassical equilibrium Eqs. (32)–(35) and we conclude that

$$r^{-\frac{3}{2}}\left\{-\frac{1}{2}\mu^{(0)}_{rrr} + \partial_\varphi \mu^{(0)}_{\varphi rr} + \mu^{(0)}_{rrr} - 2\mu^{(0)}_{\varphi r\varphi}\right\} + \ldots = 0, \tag{213}$$

$$r^{-\frac{3}{2}}\left\{-\frac{1}{2}\mu^{(0)}_{r\varphi\varphi} + \partial_\varphi \mu^{(0)}_{\varphi\varphi\varphi} + \mu^{(0)}_{r\varphi\varphi} + 2\mu^{(0)}_{\varphi r\varphi}\right\} + \ldots = 0, \tag{214}$$

$$r^{-\frac{3}{2}}\left\{-\frac{1}{2}\mu^{(0)}_{rzz} + \partial_\varphi \mu^{(0)}_{\varphi zz} + \mu^{(0)}_{rzz}\right\} + \ldots = 0, \tag{215}$$

$$r^{-\frac{3}{2}}\left\{-\frac{1}{2}\mu^{(0)}_{rr\varphi} + \partial_\varphi \mu^{(0)}_{\varphi r\varphi} + \mu^{(0)}_{rr\varphi} - \mu^{(0)}_{\varphi\varphi\varphi} + \mu^{(0)}_{\varphi rr}\right\} + \ldots = 0. \tag{216}$$

Equating to zero the coefficients of power $r^{-\frac{3}{2}}$ leads to

$$2\partial_\varphi \mu^{(0)}_{\varphi rr} + \mu^{(0)}_{rrr} - 4\mu^{(0)}_{\varphi r\varphi} = 0, \tag{217}$$

$$2\partial_\varphi \mu^{(0)}_{\varphi\varphi\varphi} + \mu^{(0)}_{r\varphi\varphi} + 4\mu^{(0)}_{\varphi r\varphi} = 0, \tag{218}$$

$$2\partial_\varphi \mu^{(0)}_{\varphi r\varphi} + \mu^{(0)}_{rr\varphi} - 2\mu^{(0)}_{\varphi\varphi\varphi} + 2\mu^{(0)}_{\varphi rr} = 0, \tag{219}$$

and

$$2\partial_\varphi \mu^{(0)}_{\varphi zz} + \mu^{(0)}_{rzz} = 0. \tag{220}$$

The last equation will not be considered further, for it can be established from Eqs. (217) and (218). To see this, we recall Eqs. (9) and (13) to recast Eq. (220) equivalently in the form

$$2\partial_\varphi \mu^{(0)}_{\varphi rr} + 2\partial_\varphi \mu^{(0)}_{\varphi\varphi\varphi} + \mu^{(0)}_{rrr} + \mu^{(0)}_{r\varphi\varphi} = 0. \tag{221}$$

But this equations can also be obtained by adding up Eqs. (217) and (218).

4.2 Nonclassical compatibility conditions

We insert the asymptotic expansion (212) into the nonclassical compatibility conditions (42)–(44) and collect terms of like powers of r, to get

$$r^{-\frac{1}{2}}\left\{\partial_\varphi\mu^{(0)}_{rr\varphi}-\mu^{(0)}_{\varphi r\varphi}+\frac{1}{2}\mu^{(0)}_{\varphi r\varphi}+\mu^{(0)}_{rrr}-\mu^{(0)}_{r\varphi\varphi}\right\}+\ldots=0, \tag{222}$$

$$r^{-\frac{1}{2}}\left\{\partial_\varphi\mu^{(0)}_{r\varphi\varphi}+\partial_\varphi\mu^{(0)}_{rrr}-\mu^{(0)}_{\varphi\varphi\varphi}-\mu^{(0)}_{\varphi rr}+\frac{1}{2}\mu^{(0)}_{\varphi\varphi\varphi}+\frac{1}{2}\mu^{(0)}_{\varphi rr}\right\}+\ldots=0, \tag{223}$$

$$r^{-\frac{1}{2}}\left\{\partial_\varphi\mu^{(0)}_{r\varphi\varphi}-\partial_\varphi\mu^{(0)}_{rrr}-\mu^{(0)}_{\varphi\varphi\varphi}+\mu^{(0)}_{\varphi rr}+\frac{1}{2}\mu^{(0)}_{\varphi\varphi\varphi}-\frac{1}{2}\mu^{(0)}_{\varphi rr}+4\mu^{(0)}_{rr\varphi}\right\}+\ldots=0. \tag{224}$$

Again, equating to zero the coefficients of power $r^{-\frac{1}{2}}$ leads to

$$\partial_\varphi\mu^{(0)}_{rr\varphi}-\frac{1}{2}\mu^{(0)}_{\varphi r\varphi}+\mu^{(0)}_{rrr}-\mu^{(0)}_{r\varphi\varphi}=0, \tag{225}$$

$$\partial_\varphi\mu^{(0)}_{r\varphi\varphi}+\partial_\varphi\mu^{(0)}_{rrr}-\frac{1}{2}\mu^{(0)}_{\varphi\varphi\varphi}-\frac{1}{2}\mu^{(0)}_{\varphi rr}=0, \tag{226}$$

$$\partial_\varphi\mu^{(0)}_{r\varphi\varphi}-\partial_\varphi\mu^{(0)}_{rrr}-\frac{1}{2}\mu^{(0)}_{\varphi\varphi\varphi}+\frac{1}{2}\mu^{(0)}_{\varphi rr}+4\mu^{(0)}_{rr\varphi}=0. \tag{227}$$

4.3 Determination of $\mu^{(0)}_{\alpha\beta\gamma}$

Eqs. (217)–(219) and (225)–(227) are 6 differential equations for the 6 unknowns $\mu^{(0)}_{rrr}$, $\mu^{(0)}_{r\varphi\varphi}$, $\mu^{(0)}_{rr\varphi}$, $\mu^{(0)}_{\varphi rr}$, $\mu^{(0)}_{\varphi\varphi\varphi}$ and $\mu^{(0)}_{\varphi r\varphi}$. The required boundary conditions can be verified to be [cf. Eq. (47)].

$$\left[\mu^{(0)}_{\varphi rr}\right]_{\varphi=\pm\pi}=\left[\mu^{(0)}_{\varphi\varphi\varphi}\right]_{\varphi=\pm\pi}=\left[\mu^{(0)}_{\varphi r\varphi}\right]_{\varphi=\pm\pi}=0. \tag{228}$$

It can be shown (cf. A) that the solutions are given by

$$\mu^{(0)}_{rrr}=\frac{\overline{B}}{2}\sin\frac{\varphi}{2}+\overline{C}\left(\sin\frac{3\varphi}{2}-\sin\frac{5\varphi}{2}\right)-\overline{A}\left(\cos\frac{3\varphi}{2}-\cos\frac{5\varphi}{2}\right), \tag{229}$$

$$\mu^{(0)}_{r\varphi\varphi}=\frac{\overline{B}}{2}\sin\frac{\varphi}{2}-\overline{C}\left(\sin\frac{3\varphi}{2}-\sin\frac{5\varphi}{2}\right)+\overline{A}\left(\cos\frac{3\varphi}{2}-\cos\frac{5\varphi}{2}\right), \tag{230}$$

$$\mu^{(0)}_{rr\varphi}=\overline{C}\left(\cos\frac{3\varphi}{2}-\cos\frac{5\varphi}{2}\right)+\overline{A}\left(\sin\frac{3\varphi}{2}-\sin\frac{5\varphi}{2}\right), \tag{231}$$

$$\mu^{(0)}_{\varphi rr}=\frac{\overline{B}}{2}\cos\frac{\varphi}{2}-\overline{C}\left(\cos\frac{3\varphi}{2}+\cos\frac{5\varphi}{2}\right)-\overline{A}\left(\sin\frac{3\varphi}{2}+\sin\frac{5\varphi}{2}\right), \tag{232}$$

$$\mu^{(0)}_{\varphi\varphi\varphi}=\frac{\overline{B}}{2}\cos\frac{\varphi}{2}+\overline{C}\left(\cos\frac{3\varphi}{2}+\cos\frac{5\varphi}{2}\right)+\overline{A}\left(\sin\frac{3\varphi}{2}+\sin\frac{5\varphi}{2}\right), \tag{233}$$

$$\mu^{(0)}_{\varphi r\varphi}=\overline{C}\left(\sin\frac{3\varphi}{2}+\sin\frac{5\varphi}{2}\right)-\overline{A}\left(\cos\frac{3\varphi}{2}+\cos\frac{5\varphi}{2}\right). \tag{234}$$

If we define

$$\overline{A}:=(c_2-c_1)\mu C^{(0)}_{I},\quad \overline{B}:=(c_2-c_1)(\lambda+\mu)B^{(0)}_{II}, \tag{235}$$

$$\overline{C}:=-(c_2-c_1)\mu E^{(0)}_{II} \tag{236}$$

then these are nothing more but the solutions for $\mu^{(0)}_{\alpha\beta\gamma}$ of Section 3.9.

5. Discussion of the asymptotic solutions

As suggested in Section 3.5, it is common to represent the leading terms of the asymptotic expansion of stresses by introducing stress intensity factors and angular functions. For the Cauchy stress, this is indicated in Eq. (108). Eqs. (108)–(110) also reveal that

$$\frac{\tilde{K}_I}{\sqrt{2\pi}} = \left[\Sigma^{(0)}_{\varphi\varphi}\right]_{\varphi=0}, \quad \frac{\tilde{K}_{II}}{\sqrt{2\pi}} = \left[\Sigma^{(0)}_{r\varphi}\right]_{\varphi=0}. \tag{237}$$

To accomplish a representation for $\mu^{(0)}_{\alpha\beta\gamma}$ similar to the one for $\Sigma^{(0)}_{\alpha\beta}$ in Eq. (108), we remark that there is only one unknown constant for Mode-I, namely $C^{(0)}_I$, but there are two unknown constants for Mode-II, $B^{(0)}_{II}$ and $E^{(0)}_{II}$ [cf. Eqs. (195)–(202)]. Therefore, in analogy to Eq. (108), we set

$$\mu^{(0)}_{\alpha\beta\gamma} = \frac{\tilde{L}_I}{\sqrt{2\pi}} g^I_{\alpha\beta\gamma}(\varphi) + \frac{\tilde{L}_{II,1}}{\sqrt{2\pi}} g^{II,1}_{\alpha\beta\gamma}(\varphi) + \frac{\tilde{L}_{II,2}}{\sqrt{2\pi}} g^{II,2}_{\alpha\beta\gamma}(\varphi), \tag{238}$$

and define for Mode–I (cf. Eq. (202))

$$\frac{\tilde{L}_I}{\sqrt{2\pi}} := \left[\mu^{(0)}_{\varphi r\varphi}\right]_{\varphi=0} = -(c_2 - c_1) 2\mu C^{(0)}_I, \tag{239}$$

rendering $\left[g^I_{\varphi r\varphi}\right]_{\varphi=0}$ to be normalized,

$$\left[g^I_{\varphi r\varphi}\right]_{\varphi=0} = 1. \tag{240}$$

To define $\tilde{L}_{II,1}$ and $\tilde{L}_{II,2}$ unambiguously, we note that $B^{(0)}_{II}$ can be determined by adding Eqs. (199) and (200) while taking $\varphi = 0$. Similarly, $E^{(0)}_{II}$ can be determined by substracting Eqs. (199) and (200) from each other while taking $\varphi = 0$. We intend to normalize the angular functions $g^{II,1}_{\alpha\beta\gamma}$ and $g^{II,2}_{\alpha\beta\gamma}$ by

$$\left[g^{II,1}_{\varphi rr}\right]_{\varphi=0} = \left[g^{II,2}_{\varphi rr}\right]_{\varphi=0} = 1, \tag{241}$$

and therefore define the stress intensity factors $\tilde{L}_{II,1}$ and $\tilde{L}_{II,2}$ by (cf. Eqs. (199) and (200))

$$\frac{\tilde{L}_{II,1}}{\sqrt{2\pi}} := \frac{1}{2}\left[\mu^{(0)}_{\varphi rr} + \mu^{(0)}_{\varphi\varphi\varphi}\right]_{\varphi=0} = \frac{1}{2}(c_2 - c_1)(\lambda + \mu) B^{(0)}_{II}, \tag{242}$$

$$\frac{\tilde{L}_{II,2}}{\sqrt{2\pi}} := \frac{1}{2}\left[\mu^{(0)}_{\varphi rr} - \mu^{(0)}_{\varphi\varphi\varphi}\right]_{\varphi=0} = (c_2 - c_1) 2\mu E^{(0)}_{II}. \tag{243}$$

The angular functions will be determined by comparison of Eqs. (238)–(243) with Eqs. (195), (196), (198)–(200) ,and (202). Explicitely, we find that

$$\begin{pmatrix} g^I_{rrr} \\ g^I_{r\varphi\varphi} \\ g^I_{rr\varphi} \\ g^I_{\varphi rr} \\ g^I_{\varphi\varphi\varphi} \\ g^I_{\varphi r\varphi} \end{pmatrix} = \frac{1}{2} \begin{pmatrix} \cos\frac{3\varphi}{2} - \cos\frac{5\varphi}{2} \\ -\cos\frac{3\varphi}{2} + \cos\frac{5\varphi}{2} \\ -\sin\frac{3\varphi}{2} + \sin\frac{5\varphi}{2} \\ \sin\frac{3\varphi}{2} + \sin\frac{5\varphi}{2} \\ -\sin\frac{3\varphi}{2} - \sin\frac{5\varphi}{2} \\ \cos\frac{3\varphi}{2} + \cos\frac{5\varphi}{2} \end{pmatrix}, \tag{244}$$

$$\begin{pmatrix} g^{II,1}_{rrr} \\ g^{II,1}_{r\varphi\varphi} \\ g^{II,1}_{rr\varphi} \\ g^{II,1}_{\varphi rr} \\ g^{II,1}_{\varphi\varphi\varphi} \\ g^{II,1}_{\varphi r\varphi} \end{pmatrix} = \begin{pmatrix} \sin\frac{\varphi}{2} \\ \sin\frac{\varphi}{2} \\ 0 \\ \cos\frac{\varphi}{2} \\ \cos\frac{\varphi}{2} \\ 0 \end{pmatrix}, \quad \begin{pmatrix} g^{II,2}_{rrr} \\ g^{II,2}_{r\varphi\varphi} \\ g^{II,2}_{rr\varphi} \\ g^{II,2}_{\varphi rr} \\ g^{II,2}_{\varphi\varphi\varphi} \\ g^{II,2}_{\varphi r\varphi} \end{pmatrix} = \frac{1}{2} \begin{pmatrix} -\sin\frac{3\varphi}{2} + \sin\frac{5\varphi}{2} \\ \sin\frac{3\varphi}{2} - \sin\frac{5\varphi}{2} \\ -\cos\frac{3\varphi}{2} + \cos\frac{5\varphi}{2} \\ \cos\frac{3\varphi}{2} + \cos\frac{5\varphi}{2} \\ -\cos\frac{3\varphi}{2} - \cos\frac{5\varphi}{2} \\ -\sin\frac{3\varphi}{2} - \sin\frac{5\varphi}{2} \end{pmatrix}. \tag{245}$$

Some comments addressing Mode-I and Mode-II crack problems are in order at this stage. In classical elasticity, there are two intensity factors in the expansion of the Cauchy stress, one for each mode. In micropolar elasticity (see, e. g., Diegele et al. [15]), there are also two stress intensity factors in the expansion of the Cauchy stress and in addition two nonclassical intensity factors in the expansion of the couple stress, one for each mode. In the present case of microstrain elasticity, there are also two stress intensity factors in the expansion of the Cauchy stress, one for each mode. However, in the expansions of the double stress there is one intensity factor for Mode-I, but there are two intensity factors for Mode-II. Actually, there are no further conditions to relate $\tilde{L}_{II,1}$ and $\tilde{L}_{II,1}$ and the numerical simulations in Part III confirm this fact.

It is also convenient to replace the constants $A_I^{(1)}, D_I^{(1)}, E_{II}^{(1)}$ and $F_{II}^{(1)}$ by the definitions

$$\tilde{l}_{I,1} := (c_2 - c_1)(\lambda + \mu) A_I^{(1)}, \tag{246}$$

$$\tilde{l}_{I,2} := -(c_2 - c_1) 2\mu D_I^{(1)}, \tag{247}$$

$$\tilde{l}_{II,1} := -(c_2 - c_1) 2\mu E_{II}^{(1)}, \tag{248}$$

$$\tilde{l}_{II,2} := -(c_2 - c_1) 2\mu F_{II}^{(1)}. \tag{249}$$

Evidently, the new constants for Mode-I and Mode-II in the expansions of $\mu^{(0)}_{\alpha\beta\gamma}$ and $\mu^{(1)}_{\alpha\beta\gamma}$ can be employed to rewrite $\Psi^{(0)}_{\alpha\beta}$ and $\Psi^{(1)}_{\alpha\beta}$. In particular, we can conclude from Eqs. (160)–(162) and (239)–(243) that

$$\Psi^{(0)}_{\alpha\beta} = \frac{\tilde{L}_I}{(c_2 - c_1)\,2\mu\sqrt{2\pi}} h^I_{\alpha\beta} + \frac{\tilde{L}_{II,1}}{(c_2 - c_1)\,(\lambda + \mu)\sqrt{2\pi}} h^{II,1}_{\alpha\beta} + \frac{\tilde{L}_{II,2}}{(c_2 - c_1)\,2\mu\sqrt{2\pi}} h^{II,2}_{\alpha\beta}, \tag{250}$$

with

$$\begin{pmatrix} h^I_{rr} \\ h^I_{\varphi\varphi} \\ h^I_{r\varphi} \end{pmatrix} = \begin{pmatrix} \cos\frac{3\varphi}{2} - \cos\frac{5\varphi}{2} \\ -\cos\frac{3\varphi}{2} + \cos\frac{5\varphi}{2} \\ -\sin\frac{3\varphi}{2} + \sin\frac{5\varphi}{2} \end{pmatrix}. \tag{251}$$

Table 1 summarizes the first two terms of the asymptotic solutions of the near-tip fields. All stresses are singular with order of singularity $r^{-\frac{1}{2}}$. Especially, the terms

$$\Sigma_{\alpha\beta} = \frac{\tilde{K}_I}{\sqrt{2\pi r}} f^I_{\alpha\beta} + \frac{\tilde{K}_{II}}{\sqrt{2\pi r}} f^{II}_{\alpha\beta} + \Sigma^{(1)}_{\alpha\beta} + \ldots, \tag{252}$$

$$\begin{pmatrix} \Sigma_{rr} \\ \Sigma_{\varphi\varphi} \\ \Sigma_{r\varphi} \end{pmatrix} = \frac{\tilde{K}_I}{4\sqrt{2\pi r}} \begin{pmatrix} 5\cos\frac{\varphi}{2} - \cos\left(\frac{3}{2}\varphi\right) \\ 3\cos\frac{\varphi}{2} + \cos\left(\frac{3}{2}\varphi\right) \\ \sin\frac{\varphi}{2} + \sin\left(\frac{3}{2}\varphi\right) \end{pmatrix} + \frac{\tilde{K}_{II}}{4\sqrt{2\pi r}} \begin{pmatrix} -5\sin\frac{\varphi}{2} + 3\sin\left(\frac{3}{2}\varphi\right) \\ -3\sin\frac{\varphi}{2} - 3\sin\left(\frac{3}{2}\varphi\right) \\ \cos\frac{\varphi}{2} + 3\cos\left(\frac{3}{2}\varphi\right) \end{pmatrix} + \tilde{k}_I \begin{pmatrix} \cos^2\varphi \\ \sin^2\varphi \\ -\frac{1}{2}\sin(2\varphi) \end{pmatrix} + \ldots. \tag{253}$$

$$\Psi_{\alpha\beta} = \overline{\Psi}_{\alpha\beta} + \sqrt{\frac{r}{2\pi}} \frac{\tilde{L}_I}{(c_2 - c_1)\,2\mu} h^I_{\alpha\beta} + \sqrt{\frac{r}{2\pi}} \frac{\tilde{L}_{II,1}}{(c_2 - c_1)\,(\lambda + \mu)} h^{II,1}_{\alpha\beta} + \sqrt{\frac{r}{2\pi}} \frac{\tilde{L}_{II,2}}{(c_2 - c_1)\,2\mu} h^{II,2}_{\alpha\beta} + r\Psi^{(1)}_{\alpha\beta} + \ldots, \tag{254}$$

$$\begin{pmatrix} \Psi_{rr} - \overline{\Psi}_{rr} \\ \Psi_{\varphi\varphi} - \overline{\Psi}_{\varphi\varphi} \\ \Psi_{r\varphi} - \overline{\Psi}_{r\varphi} \end{pmatrix} = \sqrt{\frac{r}{2\pi}} \frac{\tilde{L}_I}{(c_2 - c_1)\,2\mu} \begin{pmatrix} \cos\frac{3\varphi}{2} - \cos\frac{5\varphi}{2} \\ -\cos\frac{3\varphi}{2} + \cos\frac{5\varphi}{2} \\ -\sin\frac{3\varphi}{2} + \sin\frac{5\varphi}{2} \end{pmatrix}$$

$$+\sqrt{\frac{r}{2\pi}}\frac{\tilde{L}_{II,1}}{(c_2-c_1)(\lambda+\mu)}\begin{pmatrix}\sin\frac{\varphi}{2}\\ \sin\frac{\varphi}{2}\\ 0\end{pmatrix}$$

$$+\sqrt{\frac{r}{2\pi}}\frac{\tilde{L}_{II,2}}{(c_2-c_1)2\mu}\begin{pmatrix}-\sin\frac{3\varphi}{2}+\sin\frac{5\varphi}{2}\\ \sin\frac{3\varphi}{2}-\sin\frac{5\varphi}{2}\\ -\cos\frac{3\varphi}{2}+\cos\frac{5\varphi}{2}\end{pmatrix}$$

$$+r\frac{\tilde{l}_{I,1}}{2(c_2-c_1)(\lambda+\mu)}\begin{pmatrix}\cos\varphi\\ \cos\varphi\\ 0\end{pmatrix} \tag{255}$$

$$+r\frac{\tilde{l}_{I,2}}{(c_2-c_1)2\mu}\begin{pmatrix}\cos\varphi+\cos3\varphi\\ -\cos\varphi-\cos3\varphi\\ -\sin\varphi-\sin3\varphi\end{pmatrix}$$

$$+r\frac{\tilde{l}_{II,1}}{(c_2-c_1)2\mu}\begin{pmatrix}\sin\varphi\\ -\sin\varphi\\ \frac{1}{2}+\cos\varphi\end{pmatrix}$$

$$+r\frac{\tilde{l}_{II,2}}{(c_2-c_1)2\mu}\begin{pmatrix}\sin3\varphi\\ -\sin3\varphi\\ \frac{1}{2}-\cos3\varphi\end{pmatrix}+\ldots .$$

$$\mu_{\alpha\beta\gamma}=\frac{\tilde{L}_I}{\sqrt{2\pi r}}g^{I}_{\alpha\beta\gamma}+\frac{\tilde{L}_{II,1}}{\sqrt{2\pi r}}g^{II,1}_{\alpha\beta\gamma}+\frac{\tilde{L}_{II,2}}{\sqrt{2\pi r}}g^{II,2}_{\alpha\beta\gamma}+\mu^{(1)}_{\alpha\beta\gamma}+\ldots, \tag{256}$$

$$\begin{pmatrix}\mu_{rrr}\\ \mu_{r\varphi\varphi}\\ \mu_{rr\varphi}\\ \mu_{\varphi rr}\\ \mu_{\varphi\varphi\varphi}\\ \mu_{\varphi r\varphi}\end{pmatrix}=\frac{\tilde{L}_I}{\sqrt{2\pi r}}\frac{1}{2}\begin{pmatrix}\cos\frac{3\varphi}{2}-\cos\frac{5\varphi}{2}\\ -\cos\frac{3\varphi}{2}+\cos\frac{5\varphi}{2}\\ -\sin\frac{3\varphi}{2}+\sin\frac{5\varphi}{2}\\ \sin\frac{3\varphi}{2}+\sin\frac{5\varphi}{2}\\ -\sin\frac{3\varphi}{2}-\sin\frac{5\varphi}{2}\\ \cos\frac{3\varphi}{2}+\cos\frac{5\varphi}{2}\end{pmatrix}$$

$$+\frac{\tilde{L}_{II,1}}{\sqrt{2\pi r}}\begin{pmatrix}\sin\frac{\varphi}{2}\\ \sin\frac{\varphi}{2}\\ 0\\ \cos\frac{\varphi}{2}\\ \cos\frac{\varphi}{2}\\ 0\end{pmatrix}+\frac{\tilde{L}_{II,2}}{\sqrt{2\pi r}}\frac{1}{2}\begin{pmatrix}-\sin\frac{3\varphi}{2}+\sin\frac{5\varphi}{2}\\ \sin\frac{3\varphi}{2}-\sin\frac{5\varphi}{2}\\ -\cos\frac{3\varphi}{2}+\cos\frac{5\varphi}{2}\\ \cos\frac{3\varphi}{2}+\cos\frac{5\varphi}{2}\\ -\cos\frac{3\varphi}{2}-\cos\frac{5\varphi}{2}\\ -\sin\frac{3\varphi}{2}-\sin\frac{5\varphi}{2}\end{pmatrix}$$

$$+\tilde{l}_{I,1}\begin{pmatrix}\cos\varphi\\ \cos\varphi\\ 0\\ -\sin\varphi\\ -\sin\varphi\\ 0\end{pmatrix}+\tilde{l}_{I,2}\begin{pmatrix}\cos\varphi+\cos 3\varphi\\ -\cos\varphi-\cos 3\varphi\\ -\sin\varphi-\sin 3\varphi\\ \sin\varphi-\sin 3\varphi\\ -\sin\varphi+\sin 3\varphi\\ \cos\varphi-\cos 3\varphi\end{pmatrix}$$

$$+\tilde{l}_{II,1}\begin{pmatrix}\sin\varphi\\ -\sin\varphi\\ \frac{1}{2}+\cos\varphi\\ -1-\cos\varphi\\ 1+\cos\varphi\\ \sin\varphi\end{pmatrix}+\tilde{l}_{II,2}\begin{pmatrix}\sin 3\varphi\\ -\sin 3\varphi\\ \frac{1}{2}-\cos 3\varphi\\ 1+\cos 3\varphi\\ -1-\cos 3\varphi\\ -\sin 3\varphi\end{pmatrix}+\ldots, \tag{257}$$

$$\begin{pmatrix}\varepsilon_{rr}\\ \varepsilon_{\varphi\varphi}\\ \varepsilon_{r\varphi}\end{pmatrix}=\frac{\tilde{K}_I}{\sqrt{2\pi r}}\frac{c_1}{8\mu c_2}\begin{pmatrix}(5-8\nu)\cos\frac{\varphi}{2}-\cos\frac{3\varphi}{2}\\ (3-8\nu)\cos\frac{\varphi}{2}+\cos\frac{3\varphi}{2}\\ \sin\frac{\varphi}{2}+\sin\frac{3\varphi}{2}\end{pmatrix}$$

$$+\frac{\tilde{K}_{II}}{\sqrt{2\pi r}}\frac{c_1}{8\mu c_2}\begin{pmatrix}-(5-8\nu)\sin\frac{\varphi}{2}+3\sin\frac{3\varphi}{2}\\ -(3-8\nu)\sin\frac{\varphi}{2}-3\sin\frac{3\varphi}{2}\\ \cos\frac{\varphi}{2}+3\cos\frac{3\varphi}{2}\end{pmatrix}$$

$$+\tilde{k}^{\vec{\varepsilon}}_{I,1}\begin{pmatrix}1\\ 1\\ 0\end{pmatrix}+\tilde{k}^{\vec{\varepsilon}}_{I,2}\begin{pmatrix}\cos 2\varphi\\ -\cos 2\varphi\\ -\sin 2\varphi\end{pmatrix}+\tilde{k}^{\vec{\varepsilon}}_{II}\begin{pmatrix}\sin 2\varphi\\ -\sin 2\varphi\\ \cos 2\varphi\end{pmatrix}+\ldots, \tag{258}$$

$$\begin{pmatrix} u_r \\ u_\varphi \end{pmatrix} = \sqrt{\frac{r}{2\pi}} \frac{c_1 \tilde{K}_I}{4\mu c_2} \begin{pmatrix} (5-8\nu)\cos\frac{\varphi}{2} - \cos\frac{3\varphi}{2} \\ -(7-8\nu)\sin\frac{\varphi}{2} + \sin\frac{3\varphi}{2} \end{pmatrix}$$

$$+\sqrt{\frac{r}{2\pi}} \frac{c_1 \tilde{K}_{II}}{4c_2 \mu} \begin{pmatrix} -(5-8\nu)\sin\frac{\varphi}{2} + 3\sin\frac{3\varphi}{2} \\ -(7-8\nu)\cos\frac{\varphi}{2} + 3\cos\frac{3\varphi}{2} \end{pmatrix}$$

$$+r\left[\tilde{k}^{\vec{\varepsilon}}_{I,1}\begin{pmatrix} 1 \\ 0 \end{pmatrix} + \tilde{k}^{\vec{\varepsilon}}_{I,2}\begin{pmatrix} \cos 2\varphi \\ -\sin 2\varphi \end{pmatrix} + \tilde{k}^{\vec{\varepsilon}}_{II}\begin{pmatrix} \sin 2\varphi \\ \cos 2\varphi \end{pmatrix}\right] \qquad (259)$$
$$+\ldots.$$

Table 1.
Analytical solutions of the fields.

$\Sigma^{(0)}_{\alpha\beta}$ and $\Sigma^{(1)}_{\alpha\beta}$ are identical to those of classical elasticity. However, the terms $\varepsilon^{(0)}_{\alpha\beta}$ and $\varepsilon^{(1)}_{\alpha\beta}$ are different from the corresponding terms of classical elasticity. In particular, $\varepsilon^{(1)}_{\alpha\beta}$ includes terms arising from $\overline{\Psi}_{\alpha\beta}$. There are also qualititative differences to micropolar elasticity. For instance, terms of couple stresses corresponding to $\mu^{(0)}_{\alpha\beta\gamma}$ in Mode-II and to $\mu^{(1)}_{\alpha\beta\gamma}$ in Mode-I do not exist.

6. Concluding remarks

Closed form analytical solutions, predicted by the 3-PG-Model for Mode-I and Mode-II crack problems, have been developed in the present paper. The solutions are based on asymptotic expansions of Williams' type of the near-tip fields. The main conclusions, which can be drawn on the basis of the preceding developements, can be briefly stated as follows.

1. The first two terms in the asymptotic expansion of the components of the Cauchy stress are identical to the ones of classical elasticity. In particular, the Cauchy stress is singular with order of singularity $r^{-\frac{1}{2}}$.

2. This is in contrast to statements in other works, which rely upon boundary conditions different from the ones adopted here.

3. There are, however, significant differences in comparison to classical elasticity, in what concerns the components of macrostrain and macrodisplacement.

4. There are also significant qualitative differences in comparison to micropolar elasticity concerning the nonclassical stresses.

5. For instance, the leading terms of the double stress of Mode-II problems include two different stress intensity factors. This is a remarkable feature of the 3-PG-Model.

Appendix

In order to make the present work self-contained, we sketch briefly how to ascertain the solutions (107)–(111) from Eqs. (95)–(102). We start with the system of differential Eqs. (95)–(97), which can be proved to posses the solutions

$$\Sigma_{rr}^{(0)} = -C\cos([p+1]\varphi) - D\sin([p+1]\varphi) + \frac{3-p}{4}A\cos([p-1]\varphi) + \frac{3-p}{4}B\sin([p-1]\varphi)\ , \tag{A1}$$

$$\Sigma_{\varphi\varphi}^{(0)} = C\cos([p+1]\varphi) + D\sin([p+1]\varphi) + \frac{p+1}{4}A\cos([p-1]\varphi) + \frac{p+1}{4}B\sin([p-1]\varphi)\ , \tag{A2}$$

$$\Sigma_{r\varphi}^{(0)} = C\sin([p+1]\varphi) - D\cos([p+1]\varphi) + \frac{p-1}{4}A\sin([p-1]\varphi) + \frac{p-1}{4}B\cos([p-1]\varphi)\ . \tag{A3}$$

Here, A, B, C and D are constants of integration. In order to determine these constants, we incorporate the solutions in the boundary conditions (98). After some manipulations, we gain the following two homogeneous systems for the constants A, B, C, and D:

$$\begin{pmatrix} 2\cos([p+1]\pi) & \frac{p-1}{2}\cos([p-1]\pi) \\ 2\sin([p+1]\pi) & \frac{p+1}{2}\sin([p-1]\pi) \end{pmatrix} \begin{pmatrix} D \\ B \end{pmatrix} = \begin{pmatrix} 0 \\ 0 \end{pmatrix}, \tag{A4}$$

$$\begin{pmatrix} 2\sin([p+1]\pi) & \frac{p-1}{2}\sin([p-1]\pi) \\ 2\cos([p+1]\pi) & \frac{p+1}{2}\cos([p-1]\pi) \end{pmatrix} \begin{pmatrix} C \\ A \end{pmatrix} = \begin{pmatrix} 0 \\ 0 \end{pmatrix}. \tag{A5}$$

The conditions for the existence of nontrivial solutions are vanishing determinants of the coefficient matrices of Eqs. (A4) and (A5). It turns out that both conditions lead to the same equation

$$2\cos(p\pi)\sin(p\pi) = \sin(2p\pi) = 0, \tag{A6}$$

which has the solutions

$$p = 0, \pm\frac{1}{2}, \pm 1, \pm\frac{3}{2}, \ldots . \tag{A7}$$

The smallest value of p compatible with the restriction (82) is $p = \frac{1}{2}$, as stated in Eq. (107). For this case, the systems (A4) and (A5) imply

$$D = -\frac{3}{8}B, \quad C = \frac{1}{8}A, \tag{A8}$$

and the solutions (A1)–(A3) become

$$\Sigma_{rr}^{(0)} = \frac{1}{8}A\left(5\cos\frac{\varphi}{2} - \cos\frac{3\varphi}{2}\right) + \frac{1}{8}B\left(-5\sin\frac{\varphi}{2} + 3\sin\frac{3\varphi}{2}\right), \tag{A9}$$

$$\Sigma_{\varphi\varphi}^{(0)} = \frac{1}{8}A\left(3\cos\frac{\varphi}{2} + \cos\frac{3\varphi}{2}\right) + \frac{1}{8}B\left(-3\sin\frac{\varphi}{2} - 3\sin\frac{3\varphi}{2}\right), \tag{A10}$$

$$\Sigma_{r\varphi}^{(0)} = \frac{1}{8}A\left(\sin\frac{\varphi}{2} + \sin\frac{3\varphi}{2}\right) + \frac{1}{8}B\left(\cos\frac{\varphi}{2} + 3\cos\frac{3\varphi}{2}\right). \tag{A11}$$

These solutions, in turn, are equivalent to those of Eqs. (108)–(109). Moreover, it can be shown that for $p = \frac{1}{2}$, the solutions of Eqs. (99)–(102) might be expressed in the form (111).

References

[1] Gutkin M, Aifantis EC. Dislocations in the theory of gradient elasticity. Scripta Materialia. 1999;**40**:559-566

[2] Forest S, Sievert R. Nonlinear microstrain theories. International journal of solids and structures. 2006; **43**(24):7224-7245. Size-dependent Mechanics of Materials

[3] Williams ML. On the stress distribution at the base of a stationary crack. Journal of Applied Mechanics. 1957;**24**:109-114

[4] Muki R, Sternberg E. The influence of couple-stresses on singular concentrations in elastic solids. ZAMP. 1965;**16**:611-648

[5] Sternberg E, Muki R. The effect of couple–stresses on the stress concentration around a crack. International Journal of Solids and Structures. 1967;**3**:69-95

[6] Bogy D, Sternberg E. The effect of couple–stresses on the corner singularities due to discontinuous loadings. International Journal of Solids and Structures. 1967;**3**:757-770

[7] Bogy D, Sternberg E. The effect of couple–stresses on the corner singularities due to an asymmetric shear loading. International Journal of Solids and Structures. 1968;**4**:159-174

[8] Xia Z, Hutchinson JW. Crack tip fields in strain in gradient plasticity. Journal of the Mechanics and Physics of Solids. 1996;**44**:1621-1648

[9] Huang Y, Zhang L, Guo TF, Hwang K. Mixed mode near–tip fields for cracks in materials with strain–gradient effects. Journal of the Mechanics and Physics of Solids. 1997; **45**:439-465

[10] Huang Y, Zhang L, Guo TF, Hwang K. Near–tip fields for cracks in materials with strain–gradient effects. In: IUTAM Symposium on Nonlinear Analysis of Fracture. Netherlands: Kluwer Academic Publishers; 1997. pp. 439-465

[11] Huang Y, Chen JY, Guo TF, Zhang L, Hwang K. Analytic and numerical studies on mode I and mode II fracture in elastic–plastic materials with strain gradient effects. International Journal of Fracture. 1999; **100**:1-27

[12] Zhang L, Huang Y, Chen JY, Hwang K. The mode III full–field solution in elastic materials with strain gradient effects. International Journal of Fracture. 1998;**92**:325-348

[13] Paul H, Sridharan K. The penny–shaped crack problem in micropolar thermoelasticity. International Journal of Engineering Science. 1980;**18**: 1431-1448

[14] Chen JY, Huang Y, Ortiz M. Fracture analysis of cellular materials: A strain gradient model. Journal of the Mechanics and Physics of Solids. 1998; **46**:789-828

[15] Diegele E, Elsässer R, Tsakmakis C. Linear micropolar elastic crack-tip fields under mixed mode loading conditions. International Journal of Fracture. 2004; **129**:309-339

[16] Altan BS, Aifantis EC. On the structure of the mode III crack–tip in gradient elasticity. Scripta Metallurgica et Materialia. 1992;**26**:319-324

[17] Altan BS, Aifantis EC. On some aspects in the special theory of gradient elasticity. Journal of the Mechanical Behavior of Materials. 1997;**8**:231-282. DOI: 10.1007/s00161-014-0406-1

[18] Ru C, Aifantis EC. A simple aproach to solve boundary-value problems in

gradient elasticity. Acta Mech. 1993;**101**: 59-68

[19] Unger D, Aifantis EC. The asymptotic solution of gradient elasticity for mode iii. International Journal of Fracture. 1995;**71**:R27-R32

[20] Unger D, Aifantis EC. Strain gradient elasticity theory for antiplane shear cracks, part I: Oscillatory displacements. Theoretical and Applied Fracture Mechanics. 2000;**34**: 243-252

[21] Unger D, Aifantis EC. Strain gradient elasticity theory for antiplane shear cracks, part II: Monotonic displacements. Theoretical and Applied Fracture Mechanics. 2000;**34**:253-265

[22] Chen JY, Wei Y, Huang Y, Hutchinson JW, Hwang K. The crack tip fields in strain gradient plasticity: The asymptotic and numerical analyses. Engineering Fracture Mechanics. 1999; **64**:625-648

[23] Mousavi SM, Lazar M. Distributed dislocation technique for cracks based on non-singular dislocations in nonlocal elasticity of Helmholtz type. Engineering Fracture Mechanics. 2015; **136**:79-95

[24] Shi MX, Huang Y, Gao H, Hwang KC. Non-existence of separable crack tip field in mechanism-based strain gradient plasticity. International Journal of Solids and Structures. 2000; **37**(41):5995-6010

[25] Shi MX, Huang Y, Hwang KC. Fracture in a higher–order elastic continuum. Journal of the Mechanics and Physics of Solids. 2000;**48**: 2513-2538

[26] Vardoulakis I, Exadaktylos G, Aifantis EC. Gradient elasticity with surface energy: Mode–III crack problem. International Journal of Solids and Structures. 1996;**33**:4531-4559

[27] Karlis GF, Tsinopoulos SV, Polyzos D, Beskos DE. Boundary element analysis of mode I and mixed mode (I and II) crack problems of 2–D gradient elasticity. Computer Methods in Applied Mechanics and Engineering. 2007;**196**:5092-5103

[28] Karlis GF, Tsinopoulos SV, Polyzos D, Beskos DE. 2D and 3D boundary element analysis of mode–I cracks in gradient elasticity. Computer Modeling in Engineering and Sciences. 2008;**26**:189-207

[29] Georgiadis HG. The mode III crack problem in microstructured solids governed by dipolar gradient elasticity: Static and dynamic analysis. Journal of Applied Mechanics. 2003;**70**:517-530

[30] Askes H, Aifantis EC. Gradient elasticity in statics and dynamics: An overview of formulations, length scale identification procedures, finite element implementations and new results. International Journal of Solids and Structures. 2011;**48**:1962-1990

[31] Aifantis EC. Update on a class of gradient theories. Mechanics of Materials. 2003;**35**:259-280

Chapter 4

Mode-I and Mode-II Crack Tip Fields in Implicit Gradient Elasticity Based on Laplacians of Stress and Strain. Part III: Numerical Simulations

Abstract

A two-dimensional formulation of the 3-PG Model of implicit gradient elasticity has been developed in Part *I*. The predicted near-tip fields for Mode-I and Mode-II crack problems have been derived in Part *II*. It has been found that both the classical Cauchy stress and the nonclassical double stress are singular with the order of singularity $r^{-\frac{1}{2}}$. In the present chapter, the two-dimensional model formulation is implemented in a finite element code. For verification of the resulting finite element model, a square section with a circular hole subjected to displacement-controlled tension loading is considered and discussed. The main concerns of the chapter are, on the one hand, to validate the analytical solutions of Part *II*. On the other hand, the chapter aims to investigate the effect of nonclassical material parameters on the stress intensity factors.

Keywords: implicit gradient elasticity, finite elements, square section with a hole, mode-I and mode-II crack problems, stress intensity factors, angular functions

1. Introduction

Mode-I and Mode-II crack problems have been discussed analytically in Part *II* for the 3-PG Model of implicit gradient elasticity. Solutions for the near-tip fields have been obtained by employing the method of asymptotic expansion of Williams' type (see Williams [1]). It has been proved that both the classical Cauchy stress and the nonclassical double stress are singular with the order of singularity $r^{-\frac{1}{2}}$. Even more, the first two terms in the asymptotic expansion of the Cauchy stress are identical to those in the context of classical elasticity. The leading terms of the asymptotic expansions of the classical and the nonclassical stresses are represented by the so-called stress intensity factors.

The present chapter deals with numerical simulations that employ the 3-PG Model to solve crack problems of Mode-I and Mode-II types. A finite element model for plane strain is developed in the framework of a weak formulation based on the principle of virtual work. To verify the finite element formulation and implementation, a representative example is considered: a square section with a circular hole subjected to tension loading. The predicted stress concentration factors are compared with the corresponding stress concentration factors predicted by classical elasticity. The main objectives are to confirm the assumptions and the analytical results of Part *II* as well as to assess the effect of the nonclassical material parameters of the 3-PG Model on the stress intensity factors for Mode-I and Mode-II crack problems. It is perhaps of interest to remark that similar investigations for the case of micropolar elastic continua are provided in Diegele et al. [2].

The scope of the chapter is organized as follows: Section 2 gives some details about the implementation of the 3-PG Model in a finite element code. A square section with a circular hole subjected to tension loading is discussed in Section 3. The numerical simulations verify the finite element model and its ability to predict length scale effects. Further, they provide a first comparison to classical elasticity by calculating the corresponding stress intensity factors. Section 4 is devoted to an analysis of edge-cracked specimens. The analysis comprises, among others, the effect of material parameters on the stress intensity factors. Moreover, it indicates a very good agreement between the numerical and the analytical solutions of the angular functions. This confirms, a posteriori, the assumed symmetry conditions of the micro-deformation. The chapter closes with some concluding remarks in Section 5.

Throughout the chapter, the same notation as introduced in Part *I* applies.

2. Finite element formulation

The following formulations refer to three dimensions and apply especially to two-dimensional cases when conditions for plane strain are imposed. Let us consider once more the equilibrium equations (see Section 3.1 "The 3-PG Model as particular case of micro-strain elasticity" in Part I) to be solved,

$$\partial_i \Sigma_{ij} = 0, \tag{1}$$

$$\partial_i \mu_{ijk} + \sigma_{jk} = 0, \tag{2}$$

and specify the corresponding boundary conditions (see Section 4.7 "Boundary conditions" of Part I) as follows:

$$u_i = u_i^0 \text{ on } \partial V^{u_i}, \quad n_j \Sigma_{ji} = P_i^0 \text{ on } \partial V^{P_i}, \tag{3}$$

$$\Psi_{ij} = \Psi_{ij}^0 \text{ on } \partial V^{\Psi_{ij}}, \quad n_k \mu_{kij} = T_{ij}^0 \text{ on } \partial V^{T_{ij}}, \tag{4}$$

with

$$\partial V^{u_i} \cup \partial V^{P_i} = \partial V, \quad \partial V^{u_i} \cap \partial V^{P_i} = \emptyset, \tag{5}$$

$$\partial V^{\Psi_{ij}} \cup \partial V^{T_{ij}} = \partial V, \quad \partial V^{\Psi_{ij}} \cap \partial V^{T_{ij}} = \emptyset. \tag{6}$$

These equations reflect Dirichlet boundary conditions for the macro-displacement and the micro-deformation and Neumann boundary conditions for the Cauchy stress and the couple stress.

The first step toward a finite element formulation is to elaborate a weak form of the above boundary value problem.

2.1 Weak form of the boundary value problem

Let the fields $u_i(\boldsymbol{x})$ and $\Psi_{ij}(\boldsymbol{x})$ as well as the so-called virtual fields $\delta u_j(\boldsymbol{x})$ and $\delta \Psi_{jk}(\boldsymbol{x})$ belong to the following function spaces:

$$u_i \in \mathcal{S} := \{u_i \ |u_i \in \mathcal{H}^1(\overline{V}),\ u_i = u_i^0 \text{ on } \partial V^{u_i}\}, \tag{7}$$

$$\Psi_{ij} \in \mathcal{T} := \left\{\Psi_{ij} \ |\Psi_{ij} \in \mathcal{H}^1(\overline{V}),\ \Psi_{ij} = \Psi_{ij}^0 \text{ on } \partial V^{\Psi_{ij}}\right\}, \tag{8}$$

$$\delta u_j \in \mathcal{V} := \{\delta u_j \ |\delta u_j \in \mathcal{H}^1(\overline{V}),\ \delta u_j = 0 \text{ on } \partial V^{u_i}\}, \tag{9}$$

$$\delta\Psi_{jk} \in \mathcal{W} := \{\delta\Psi_{jk} \ |\delta\Psi_{jk} \in \mathcal{H}^1(\overline{V}),\ \delta\Psi_{jk} = 0 \text{ on } \partial V^{\Psi_{ij}}\}. \tag{10}$$

Now, multiply Eq. (1) by δu_j and Eq. (2) by $\delta\Psi_{jk}$ and take the integrals over V,

$$\int_V \delta u_j \partial_i \Sigma_{ij} \, \mathrm{d}v = 0, \tag{11}$$

$$\int_V \delta\Psi_{jk} \left(\partial_i \mu_{ijk} + \sigma_{jk}\right) \mathrm{d}v = 0. \tag{12}$$

Next, add up these two equations, use partial integration, the divergence theorem, and the boundary conditions (3) and (4) to receive

$$\begin{aligned} &\int_V (\partial_i \delta u_j)\, \Sigma_{ij} \, \mathrm{d}v - \int_V (\delta\Psi_{jk})\, \sigma_{jk} \, \mathrm{d}v + \int_V (\partial_i \delta\Psi_{jk})\, \mu_{ijk} \, \mathrm{d}v \\ &- \int_{\partial V^{\mathrm{P}}} (\delta u_j)\, P_j^0 \, \mathrm{d}a - \int_{\partial V^{\mathrm{T}}} (\delta\Psi_{jk})\, T_{jk}^0 \, \mathrm{d}a = 0. \end{aligned} \tag{13}$$

As usually, in favor of a short notation, we use the integral over ∂V^P to indicate the summation of single integrals over ∂V^{P_i}, which generally do not coincide. The meaning of the integration over ∂V^T is analogous. Eq. (13) is the weak form of the boundary value problem and is the starting point of the finite element formulation.

2.2 Discretization

According to the finite element method (see, e.g., Hughes [3]), the domain V is approximated by n_{el} finite elements V_e, so that

$$V \approx V^h := \bigcup_{e=1}^{n_{\mathrm{el}}} V_e. \tag{14}$$

The elements are connected to each other at selected points called nodal points, or simply nodes, and the following notation holds.

$\mathcal{K}^{\boldsymbol{u}}$: Set of global node numbers with macro-displacement degree of freedom.

$\mathcal{K}^{\boldsymbol{\Psi}}$: Set of global node numbers with micro-deformation degree of freedom.

The exact solutions u_i and Ψ_{ij} are approximated by

$$u_i(\boldsymbol{x}) \approx u_i^h(\boldsymbol{x}) := \sum_{A \in \mathcal{K}^{\boldsymbol{u}}} N_A^{\boldsymbol{u}}(\boldsymbol{x})\, u_i^A, \tag{15}$$

$$\Psi_{ij}(\boldsymbol{x}) \approx \Psi_{ij}^h(\boldsymbol{x}) := \sum_{A \in \mathcal{K}^{\boldsymbol{\Psi}}} N_A^{\Upsilon}(\boldsymbol{x})\, \Psi_{ij}^A, \tag{16}$$

where u_i^A and Ψ_{ij}^A are the unknown values of u_i and Ψ_{ij} at node A. The so-called shape functions $N_A^{\boldsymbol{u}}$ and $N_A^{\boldsymbol{\Psi}}$ belong to finite-dimensional function spaces $\mathcal{S}^h$ and $\mathcal{T}^h$, which approximate the function spaces $\mathcal{S}$ and $\mathcal{T}$, respectively. Similarly, δu_j and $\delta\Psi_{jk}$ are approximated by

$$\delta u_j(\boldsymbol{x}) \approx \delta u_j^h(\boldsymbol{x}) := \sum_{B \in \mathcal{K}^{\boldsymbol{u}}} N_B^{\boldsymbol{u}}(\boldsymbol{x})\, \delta u_j^B, \tag{17}$$

$$\delta \Psi_{jk}(\boldsymbol{x}) \approx \delta \Psi_{jk}^h(\boldsymbol{x}) := \sum_{B \in \mathcal{K}^{\boldsymbol{\Psi}}} N_B^{\boldsymbol{u}}(\boldsymbol{x})\, \delta \Psi_{jk}^B, \tag{18}$$

with δu_j^B and $\delta \Psi_{jk}^B$ being constants. The functions δu_j^h and $\delta \Psi_{jk}^h$ are elements of finite-dimensional function spaces $\mathcal{V}^h$ and $\mathcal{W}^h$, which approximate the function spaces $\mathcal{V}$ and $\mathcal{W}$, respectively.

We may use the above approximations (17) and (18) to rewrite Eq. (13) in the form

$$\begin{aligned} \sum_{e=1}^{n_{\mathrm{el}}} \Big\{ \int_{V_e} (\partial_i N_B^{\boldsymbol{u}})\, \delta u_j^B\, \Sigma_{ij}\, \mathrm{d}v - \int_{V_e} N_B^{\boldsymbol{\Psi}}\, \delta \Psi_{jk}^B\, \sigma_{jk}\, \mathrm{d}v + \int_{V_e} (\partial_i N_B^{\boldsymbol{\Psi}})\, \delta \Psi_{jk}^B\, \mu_{ijk}\, \mathrm{d}v \\ - \int_{\partial V_e^P} N_B^{\boldsymbol{u}}\, \delta u_j^B\, P_j^0\, \mathrm{d}a - \int_{\partial V_e^T} N_B^{\boldsymbol{\Psi}}\, \delta \Psi_{jk}^B\, T_{jk}^0\, \mathrm{d}a \} = 0, \end{aligned} \tag{19}$$

with the meaning of ∂V_e^P and ∂V_e^T being obvious.

We next employ the elasticity laws (see Section 3.1 "The 3-PG Model as particular case of micro-strain elasticity" of Part I)

$$\Sigma_{ij} = \frac{c_2}{c_1}\, \mathbb{C}_{ijmn}\, \varepsilon_{mn} - \frac{c_2 - c_1}{c_1}\, \mathbb{C}_{ijmn}\, \Psi_{mn}, \tag{20}$$

$$\sigma_{jk} = \frac{c_2 - c_1}{c_1}\, (\mathbb{C}_{jkmn}\, \varepsilon_{mn} - \mathbb{C}_{jkmn}\, \Psi_{mn}), \tag{21}$$

$$\mu_{ijk} = (c_2 - c_1)\, \partial_i \Psi_{mn}\, \mathbb{C}_{mnjk}, \tag{22}$$

in order to replace the stresses Σ_{ij} and σ_{jk} as well as the double stresses μ_{ijk} in Eq. (19):

$$\begin{aligned} \sum_{e=1}^{n_{\mathrm{el}}} \Big\{ \int_{V_e} (\partial_i N_B^{\boldsymbol{u}})\, \delta u_j^B \left(\frac{c_2}{c_1}\, \mathbb{C}_{ijmn}\, \partial_m u_n - \frac{c_2 - c_1}{c_1}\, \mathbb{C}_{ijmn}\, \Psi_{mn} \right) \mathrm{d}v \\ - \frac{c_2 - c_1}{c_1} \int_{V_e} N_B^{\boldsymbol{\Psi}}\, \delta \Psi_{jk}^B\, (\mathbb{C}_{jkmn}\, \partial_m u_n - \mathbb{C}_{jkmn}\, \Psi_{mn})\, \mathrm{d}v \\ + (c_2 - c_1) \int_{V_e} (\partial_i N_B^{\boldsymbol{\Psi}})\, \delta \Psi_{jk}^B\, \partial_i \Psi_{mn}\, \mathbb{C}_{mnjk}\, \mathrm{d}v \\ - \int_{\partial V_e^P} N_B^{\boldsymbol{u}}\, \delta u_j^B\, P_j^0\, \mathrm{d}a - \int_{\partial V_e^T} N_B^{\boldsymbol{\Psi}}\, \delta \Psi_{jk}^B\, T_{jk}^0\, \mathrm{d}a \} = 0. \end{aligned} \tag{23}$$

Finally, we use the approximations (15) and (16), to find

$$\begin{aligned} \sum_{e=1}^{n_{\mathrm{el}}} \Big\{ \int_{V_e} (\partial_i N_B^{\boldsymbol{u}})\, \delta u_j^B \left[\frac{c_2}{c_1}\, \mathbb{C}_{ijmn}\, (\partial_m N_A^{\boldsymbol{u}})\, u_n^A - \frac{c_2 - c_1}{c_1}\, \mathbb{C}_{ijmn}\, N_A^{\Upsilon}\, \Psi_{mn}^A \right] \mathrm{d}v \\ + \frac{c_2 - c_1}{c_1} \int_{V_e} N_B^{\boldsymbol{\Psi}}\, \delta \Psi_{jk}^B\, [\mathbb{C}_{jkmn}\, (\partial_m N_A^{\mathbf{u}})\, u_n^A - \mathbb{C}_{jkmn}\, N_A^{\boldsymbol{\Psi}}\, \Psi_{mn}^A]\, \mathrm{d}v \\ + (c_2 - c_1) \int_{V_e} (\partial_i N_B^{\boldsymbol{\Psi}})\, \delta \Psi_{jk}^B\, (\partial_i N_A^{\boldsymbol{\Psi}})\, \Psi_{mn}^A\, \mathbb{C}_{mnjk}\, \mathrm{d}v \\ - \int_{\partial V_e^P} N_B^{\boldsymbol{u}}\, \delta u_j^B\, P_j^0\, \mathrm{d}a - \int_{\partial V_e^T} N_B^{\boldsymbol{\Psi}}\, \delta \Psi_{jk}^B\, T_{jk}^0\, \mathrm{d}a \Big\} = 0, \end{aligned} \tag{24}$$

or equivalently,

$$\begin{aligned}\sum_{e=1}^{n_{\rm el}} \Bigg\{ & \delta u_j^B \left[\frac{c_2}{c_1}\int_{V_e} (\partial_i N_B^{\boldsymbol{u}})\, \mathbb{C}_{ijmn}\, (\partial_m N_A^{\boldsymbol{u}})\, \mathrm{d}v\right] u_n^A \\ & -\delta u_j^B \left[\frac{c_2-c_1}{c_1}\int_{V_e} (\partial_i N_B^{\boldsymbol{u}})\, \mathbb{C}_{ijmn}\, N_A^{\boldsymbol{\Psi}}\, \mathrm{d}v\right] \Psi_{mn}^A \\ & -\delta \Psi_{jk}^B \left[\frac{c_2-c_1}{c_1}\int_{V_e} N_B^{\boldsymbol{\Psi}}\, \mathbb{C}_{jkmn}\, (\partial_m N_A^{\boldsymbol{u}})\, \mathrm{d}v\right] u_n^A \\ & +\delta \Psi_{jk}^B \left[\frac{c_2-c_1}{c_1}\int_{V_e} N_B^{\boldsymbol{\Psi}}\, \mathbb{C}_{jkmn}\, N_A^{\boldsymbol{\Psi}}\, \mathrm{d}v\right] \Psi_{mn}^A \\ & +\delta \Psi_{jk}^B \left[(c_2-c_1)\int_{V_e} (\partial_i N_B^{\boldsymbol{\Psi}})\, (\partial_i N_A^{\boldsymbol{\Psi}})\, \mathbb{C}_{jkmn}\, \mathrm{d}v\right] \Psi_{mn}^A \\ & -\delta u_j^B \left[\int_{\partial V_e^P} N_B^{\boldsymbol{u}} P_j^0\, \mathrm{d}a\right] - \delta \Psi_{jk}^B \left[\int_{\partial V_e^T} N_B^{\boldsymbol{\Psi}} T_{jk}^0\, \mathrm{d}a\right]\Bigg\} = 0. \end{aligned} \tag{25}$$

The form of this equation suggests to define the element stiffness matrices:

$$K_{jBnA}^{\boldsymbol{uu},(e)} := \frac{c_2}{c_1}\int_{V_e} (\partial_i N_B^{\boldsymbol{u}})\, \mathbb{C}_{ijmn}\, (\partial_m N_A^{\boldsymbol{u}})\, \mathrm{d}v, \tag{26}$$

$$K_{jBmnA}^{\boldsymbol{u\Psi},(e)} := -\frac{c_2-c_1}{c_1}\int_{V_e} (\partial_i N_B^{\boldsymbol{u}})\, \mathbb{C}_{ijmn}\, N_A^{\boldsymbol{\Psi}}\, \mathrm{d}v, \tag{27}$$

$$K_{jkBnA}^{\boldsymbol{\Psi u},(e)} := -\frac{c_2-c_1}{c_1}\int_{V_e} N_B^{\boldsymbol{\Psi}}\, \mathbb{C}_{jkmn}\, (\partial_m N_A^{\boldsymbol{u}})\, \mathrm{d}v, \tag{28}$$

$$K_{jkBmnA}^{\boldsymbol{\Psi\Psi},(e)} := \frac{c_2-c_1}{c_1}\int_{V_e} \left[N_B^{\boldsymbol{\Psi}}\, \mathbb{C}_{jkmn}\, N_A^{\boldsymbol{\Psi}} + c_1\, (\partial_i N_B^{\boldsymbol{\Psi}})\, (\partial_i N_A^{\boldsymbol{\Psi}})\, \mathbb{C}_{jkmn}\right] \mathrm{d}v, \tag{29}$$

and the element force vectors:

$$F_{jB}^{\boldsymbol{u},(e)} := \int_{V_e^P} N_B^{\boldsymbol{u}} P_j^0\, \mathrm{d}a, \tag{30}$$

$$F_{jkB}^{\boldsymbol{\Psi},(e)} := \int_{V_e^T} N_B^{\boldsymbol{\Psi}} T_{jk}^0\, \mathrm{d}a, \tag{31}$$

and to recast Eq. (25) as

$$\begin{aligned}\sum_{e=1}^{n_{\rm el}} \Big\{ & \delta u_j^B \left[K_{jBnA}^{\boldsymbol{uu},(e)} u_n^A + K_{jBmnA}^{\boldsymbol{u\Psi},(e)} \Psi_{mn}^A - F_{jB}^{\boldsymbol{u},(e)}\right] \\ & +\delta \Psi_{jk}^B \left[K_{jkBnA}^{\boldsymbol{\Psi u},(e)} u_n^A + K_{jkBmnA}^{\boldsymbol{\Psi\Psi},(e)} \Psi_{mn}^A - F_{jkB}^{\boldsymbol{\Psi},(e)}\right]\Big\} = 0. \end{aligned} \tag{32}$$

As δu_j^B and $\delta \Psi_{jk}^B$ may be chosen arbitrary, we obtain the system of equations:

$$\sum_{e=1}^{n_{\rm el}} \left[K_{jBnA}^{\boldsymbol{uu},(e)} u_n^A + K_{jBmnA}^{\boldsymbol{u\Psi},(e)} \Psi_{mn}^A - F_{jb}^{\boldsymbol{u},(e)}\right] = 0, \tag{33}$$

$$\sum_{e=1}^{n_{\rm el}} \left[K_{jkBnA}^{\boldsymbol{\Psi u},(e)} u_n^A + K_{jkBmnA}^{\boldsymbol{\Psi\Psi},(e)} \Psi_{mn}^A - F_{jkb}^{\boldsymbol{\Psi},(e)}\right] = 0. \tag{34}$$

This means, that the solution of the boundary value problems (1)–(6) is reduced to the solution of the systems (33) and (34).

The proceeding approximation of the weak form, constrained to plane strain states, has been implemented in the finite element code FEAP. Isoparametric elements are employed, that is, the space coordinates are represented by using the shape functions, $N^{\boldsymbol{u}}_A$,

$$x_i \approx x_i^h := \sum_{A \in \mathcal{K}^{\boldsymbol{u}}} N^{\boldsymbol{u}}_A x_i^A, \tag{35}$$

where x_i^A are constants.

3. Square section with a circular hole

The main objective of this section is to validate the implemented finite element code. To this end, we consider the plane strain problem shown in **Figure 1a**, where the square section with a circular hole, located at the center of the section, is stretched in the y-direction. The length of the section is $b = 5\,\text{mm}$, while the radius of the hole is $r = 0.25\,\text{mm}$. With respect to the Cartesian coordinate system $\{x, y\}$, the boundaries $x = \pm\frac{b}{2}$ are assumed to be free of classical and nonclassical tractions. At the boundary $y = -\frac{b}{2}$, the macro-displacement component u_y, the component P_x of the classical traction, and the nonclassical traction components T_{ij} are assumed to vanish. At the boundary $y = \frac{b}{2}$, the macro-displacement in the y-direction is given by $u_y = 0.1$ mm, while $P_x = 0$ and $T_{ij} = 0$ are imposed. The whole circular hole is assumed to be free of classical and nonclassical tractions. For a small circular hole, a nearly uniform stress component

$$\Sigma_0 := \left[\Sigma_{yy}\right]_{y=\frac{b}{2}} \tag{36}$$

will be required to realize the given boundary conditions.

The most simple case in classical elasticity, analogous to the boundary value problem above, is to consider the square section in the context of a plane stress problem subjected to the traction boundary condition $\left[\Sigma_{yy}\right]_{y=\frac{b}{2}} = \Sigma_0$. Attention is focused on the distribution of Σ_{yy} along the section A-A (see **Figure 1a**), as a function of the local coordinate

$$a := x - r, \tag{37}$$

with $x \geq r$. A so-called stress concentration factor k is defined by

$$k := \frac{\Sigma_{yy}^*}{\Sigma_0}, \quad \Sigma_{yy}^* := \Sigma_{yy}(a = 0, y = 0), \tag{38}$$

and turns out to be $k = 3$ (see, e.g., Gould [4], p. 124).

Now consider the square section in **Figure 1a** in the context of a plane strain problem with the boundary conditions stated in the first paragraph within classical elasticity. The stress distribution along the section A-A has been determined by employing the standard elastic element of FEAP using the classical material parameters

$$E = 100 \text{ GPa}, \quad \nu = 0,3. \tag{39}$$

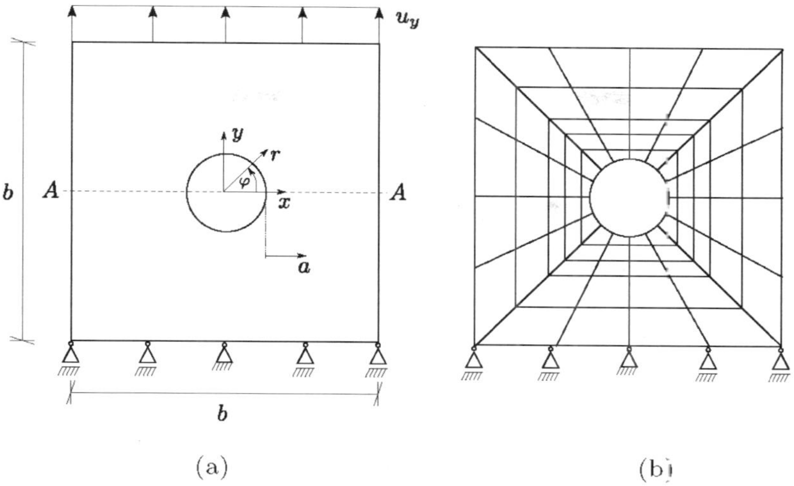

Figure 1.
Plane strain problem of a square section with a circular hole subjected to displacement-controlled tension loading. (a) Geometry and loading conditions. (b) Mesh (schematically).

As depicted schematically in **Figure 1b**, a radial mesh of 16 stripes of quadratic elements with reduced integration, that is, elements with 8 nodes, is used. Every stripe consists of 180 elements and the elements are chosen to decrease in size the closer to the circular hole the elements are placed. In the context of classical elasticity, a mesh consisting of 720 such stripes is used, whereas the mesh for the simulations of the 3-PG Model only needs 72 such stripes of elements.

Figure 2 illustrates the distribution of the dimensionless stress $\frac{\Sigma_{yy}}{\Sigma_0}$ along the section A-A. The value of $\frac{\Sigma_{yy}}{\Sigma_0}$ at $a = 0$ represents the stress concentration factor k and turns out to be $k = 3,14$ for classical elasticity.

We consider next corresponding distributions predicted by the 3-PG Model. To this end, we employ the finite element model developed in the last section. The classical material parameters are given by Eq. (39). For the purposes of the present chapter, we find it convenient to use the nonclassical parameters c_1 and

$$c_3 := \frac{c_2 - c_1}{c_1}. \tag{40}$$

Note that since $c_2 > c_1 > 0$ (see Section 3.1 "The 3-PG Model as particular case of micro-strain elasticity" of Part I), the constrain $c_3 > 0$ applies. It becomes apparent, from the elasticity laws (20)–(22), that for $c_3 \to 0$, the 3-PG Model will approach to classical elasticity. Hence, we expect the distributions of the dimensionless stress $\frac{\Sigma_{yy}}{\Sigma_0}$ along the section A-A to be close to the classical one whenever c_3 is sufficiently small. Indeed, **Figure 2** confirms this expectation: It can be seen, that for sufficiently large values of a, the two graphs almost coincide. Another way to illustrate this issue is to consider the effect of the nonclassical material parameter c_3 on the values of the stress concentration factor k. We expect that for $c_1 = \text{const.}$ and $c_3 \to 0$, the values of k will approach to the classical value $k = 3,14$. This is exactly what we can observe in **Figure 3**. Alternatively, we can consider the effect of the material parameter c_1 on the values of the stress concentration factor k. For the assumed geometry of the specimen, the imposed boundary conditions and for a fixed value $c_3 = 1$, the effect of c_1 on k is illustrated in **Figure 4** by the graph referred to as

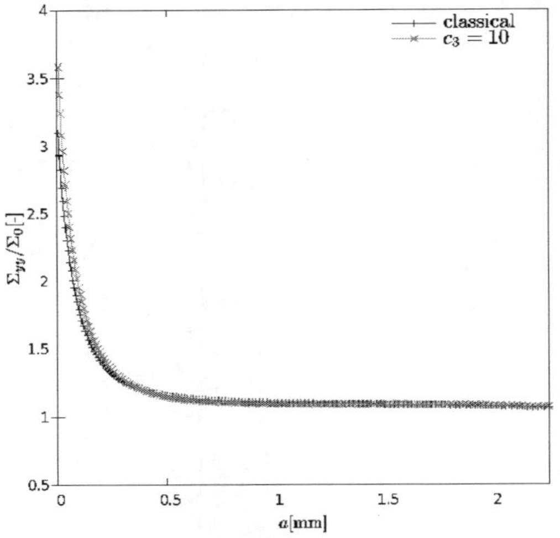

Figure 2.
Distribution of the dimensionless stress Σ_{yy}/Σ_0 along the section A-A with $E = 100\,\text{GPa}$, $\nu = 0, 3$, $c_1 = 0, 2\ \text{mm}^2$, and $c_3 = 10$.

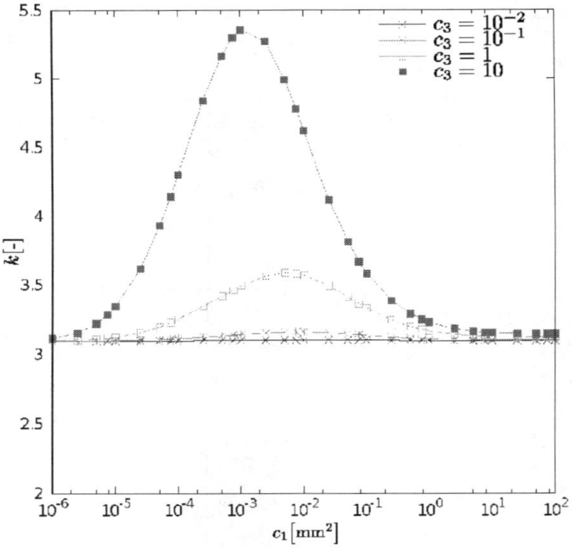

Figure 3.
Distribution of the stress concentration factor k predicted by the 3-PG model.

specimen 1. A convenient way to illustrate the effect of c_1 is to consider further boundary value problems. Thus, we consider in addition three further problems, denoted as specimen 4, 20, and 200, which arise by multiplying the given geometry of the specimen and the imposed boundary conditions with the factors $n = 4, 20$ and 200, respectively. The corresponding graphs of k as a function of c_1 are displayed in **Figure 4** and are referred to as specimens 4, 20, and 200, respectively. Keeping in mind, that $\sqrt{c_1}$ is an internal material length, we rescale the abscissa c_1 by considering the graphs of k as a function of $\frac{c_1}{n^2}$. We expect, that all distributions should coincide, and in fact this is shown in **Figure 5**.

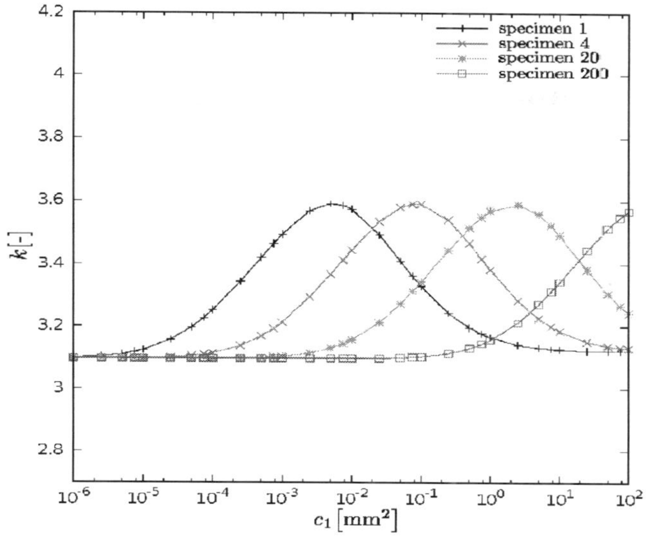

Figure 4.
Effect of the material parameter c_1 on the stress concentration factor k ($c_3 = 1$).

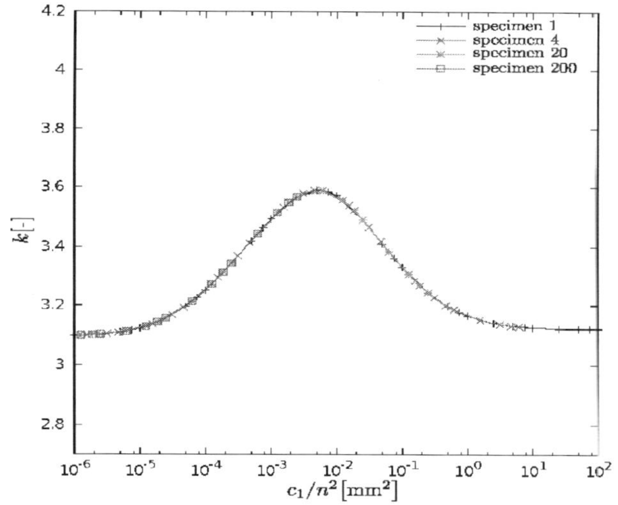

Figure 5.
Distribution of the stress concentration factor k as a function of c_1/n^2 ($c_3 = 1$).

Altogether, the calculated responses reflect the expected results, which in turn provides a validation of the developed and implemented finite element approximation of the 3-PG Model.

4. Finite element analysis of crack problems

4.1 Edge-cracked specimen

The remaining analysis is referred to the edge-cracked specimen shown in **Figure 6a**. The assumed width and length of the specimen are, respectively,

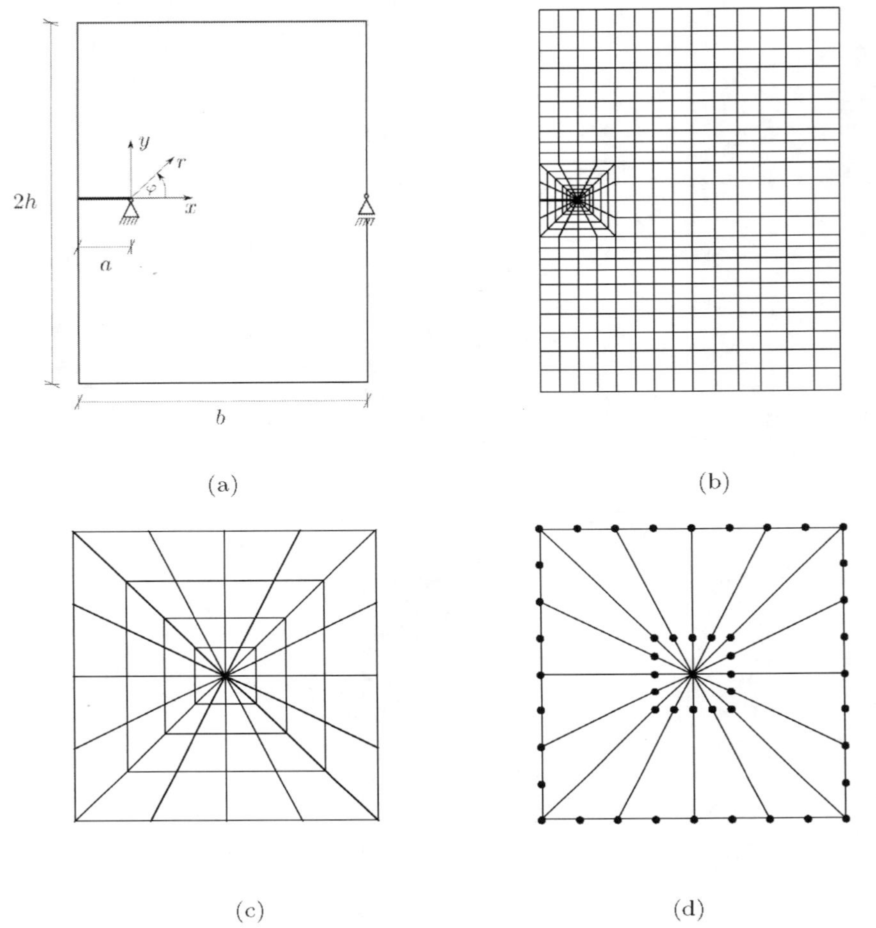

Figure 6.
Edge-cracked specimen with finite width b, length 2h and crack length a. (a) Geometry. (b) Mesh. (c) Radial mesh around crack tip. (d) Singular elements around crack tip.

$b = 11$ mm and $2h = 20$ mm, while the crack length is chosen to be $a = 1$ mm. The origin of the Cartesian and the cylindrical coordinate systems, to which we refer, is located at the crack tip.

The specimen is discretized by two meshes (see **Figure 6b**), a rectangular mesh for the main part of the specimen, consisting of 320 quadratic eight-node elements (with reduced integration) and a radial mesh around the crack tip. The radial mesh, schematically shown in **Figure 6c**, consists of 16 stripes of quadratic elements with reduced integration, that is, elements with eight nodes, are used. Every strip consists of 45 elements, which decrease in size the closer to the crack tip they are placed. The elements, which contain the crack tip itself, are singular and are the so-called quarter point elements. That means, that one edge of each quadratic element degenerates into a point and two intermediate nodes of two adjacent edges are repositioned. To be more specific, they are moved to a position only a quarter of the edge length away from the crack tip, as indicated in **Figure 6d**. Due to this repositioning, the corresponding shape functions differ from those of regular

8–node elements. Specifically, the shape functions of quarter point elements consist of linear shape functions extended by a term proportional to the square root of the corresponding component of the position vector (see Henshell and Shaw [5] or Barsoum [6]).

For all calculations in the remainder of the chapter, the values of the classical material parameters E and ν are given by Eq. (39) and the following Dirichlet boundary conditions are imposed:

$$[u_r]_{r=0} = [u_\varphi]_{r=0} = 0, \quad [u_\varphi]_{r=10\text{ mm},\varphi=0} = 0, \tag{41}$$

which excludes the possibility of rigid body motions.

The crack faces are subjected to the following boundary conditions for the classical stress: Mode-I near-tip fields are produced by imposing an internal pressure $p = 100$ MPa to act, while Mode-II near-tip fields are enforced by subjecting the crack faces to a shear stress loading of 100 MPa. All nonclassical traction components are supposed to vanish for both Mode-I and Mode-II crack types.

4.2 Stress intensity factors

Finite element simulations of Mode-I and Mode-II crack problems allow to verify, numerically, the order of singularity and the angular functions as well as to determinate the stress intensity factors. First, we verify the order of singularity and show how to determine stress intensity factors numerically. Motivated by the analytical solutions in Section 5 “Discussion of the asymptotic solutions” of Part II, assume that the stress $\boldsymbol{\Sigma}$ and the double stress $\boldsymbol{\mu}$ in the vicinity of the crack tip are given by

$$\Sigma_{\alpha\beta} = (2\pi r)^{p-1}\left[\tilde{K}_I f^I_{\alpha\beta}(\varphi) + \tilde{K}_{II} f^{II}_{\alpha\beta}(\varphi)\right], \tag{42}$$

$$\mu_{\alpha\beta\gamma} = (2\pi r)^{p-1}\left[\tilde{L}_I g^I_{\alpha\beta\gamma}(\varphi) + \tilde{L}_{II,1} g^{II,1}_{\alpha\beta\gamma}(\varphi) + \tilde{L}_{II,2} g^{II,2}_{\alpha\beta\gamma}(\varphi)\right]. \tag{43}$$

Then,

$$[\Sigma_{\varphi\varphi}]_{\varphi=0} = (2\pi r)^{p-1}\tilde{K}_I, \tag{44}$$

$$[\Sigma_{r\varphi}]_{\varphi=0} = (2\pi r)^{p-1}\tilde{K}_{II}, \tag{45}$$

$$[\mu_{\varphi r\varphi}]_{\varphi=0} = (2\pi r)^{p-1}\tilde{L}_I, \tag{46}$$

$$\frac{1}{2}\left[\mu_{\varphi rr} + \mu_{\varphi\varphi\varphi}\right]_{\varphi=0} = (2\pi r)^{p-1}\tilde{L}_{II,1}, \tag{47}$$

$$\frac{1}{2}\left[\mu_{\varphi rr} - \mu_{\varphi\varphi\varphi}\right]_{\varphi=0} = (2\pi r)^{p-1}\tilde{L}_{II,2}, \tag{48}$$

and hence

$$\log\,[\Sigma_{\varphi\varphi}]_{\varphi=0} = (p-1)\log(2\pi r) + \log\tilde{K}_I, \tag{49}$$

$$\log\,[\Sigma_{r\varphi}]_{\varphi=0} = (p-1)\log(2\pi r) + \log\tilde{K}_{II}, \tag{50}$$

$$\log\,[\mu_{\varphi r\varphi}]_{\varphi=0} = (p-1)\log(2\pi r) + \log\tilde{L}_I, \tag{51}$$

$$\log\left(\frac{1}{2}\left[\mu_{\varphi rr}+\mu_{\varphi\varphi\varphi}\right]_{\varphi=0}\right)=(p-1)\log(2\pi r)+\log\tilde{L}_{II,1},\qquad(52)$$

$$\log\left(\frac{1}{2}\left[\mu_{\varphi rr}-\mu_{\varphi\varphi\varphi}\right]_{\varphi=0}\right)=(p-1)\log(2\pi r)+\log\tilde{L}_{II,2}.\qquad(53)$$

These equations indicate, respectively, a linear respone with slope $(p-1)$. This means that, as in the case of classical elastic fracture mechanics, one can fit the exponent p on the basis of values of stress components calculated by the finite element method at the nodes ahead of the crack tip (see, e.g., **Figure 7**). Computed responses with the finite element model have confirmed the value $p=\frac{1}{2}$ with great accuracy. Therefore, the value $p=\frac{1}{2}$ will be fixed, to avoid errors owing to inaccurate numerical determination of p when discussing predicted responses. All stress intensity factors are determined from Eqs. (49)–(53) for the fixed value $p=\frac{1}{2}$ by applying the least square method. It can be recognized from **Figure 7**, that the linear response of the stress applies for a radius $r\in\left[10^{-7},10^{-1}\right]$. The numerical determination of the angular functions below is referred to a fixed radius $r=5\cdot10^{-6}$ mm. The verification of the angular functions and the effect of material parameters on the stress intensity factor is discussed seperately for Mode-I and Mode-II crack problems.

4.3 Results for mode-I

Figures 8–11 display plots of the angular functions $f^{I}_{\alpha\beta}$, $h^{I}_{\alpha\beta}$ and $g^{I}_{\alpha\beta\gamma}$ for both the analytical and the finite element solutions. The plots of the latter are constructed by dividing the values of $\Sigma_{\alpha\beta}$, $\Psi_{\alpha\beta}-\overline{\Psi}_{\alpha\beta}$, and $\mu_{\alpha\beta\gamma}$ at a radius $r=5\cdot10^{-6}$ mm by $\frac{\tilde{K}_I}{\sqrt{2\pi r}}$, $\frac{r}{2\pi}\frac{\tilde{L}_I}{(c_2-c_1)2\mu}=\frac{r}{2\pi}\frac{\tilde{L}_I}{c_1c_3 2\mu}$ and $\frac{\tilde{L}_I}{\sqrt{2\pi r}}$, respectively. The stress intensity factors $\tilde{K}_I$ and $\tilde{L}_I$ have been determined by least square fitting as described in the last section. In the finite element computations, the values $c_1=10^{-4}\,\text{mm}^2$ and $c_3=1$ for the nonclassical material parameters are chosen. The values of $\overline{\Psi}_{ij}$ are determined to be

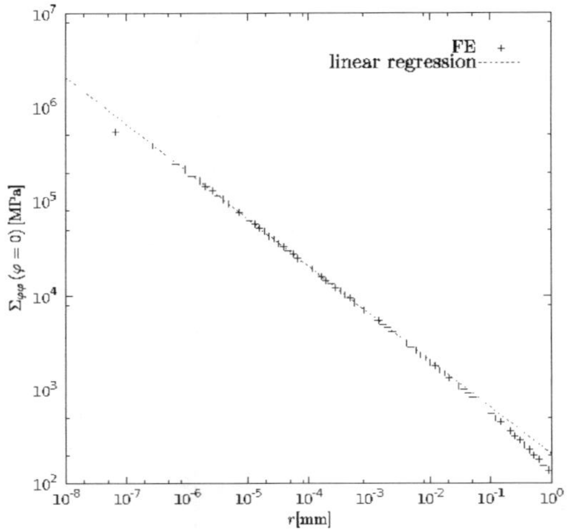

Figure 7.
Least square fitting of the distribution of $\left[\Sigma_{\varphi\varphi}\right]_{\varphi=0}$ computed by the finite element model in the vicinity of the crack tip.

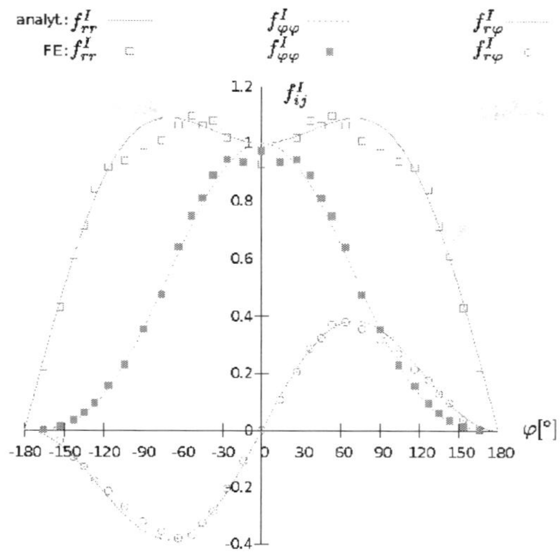

Figure 8.
Graphs of the angular functions $f^I_{\alpha\beta}(\varphi)$ of mode-I for $\varphi \in [-180°, 180°]$.

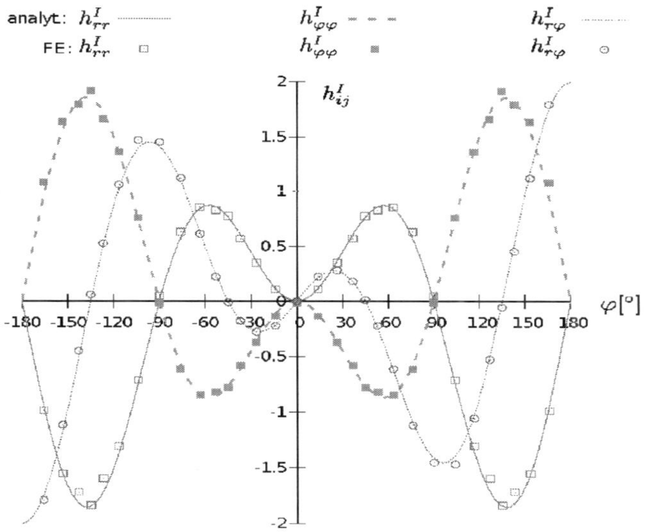

Figure 9.
Graphs of the angular functions $h^I_{\alpha\beta}(\varphi)$ of mode-I for $\varphi \in [-180°, 180°]$.

$\overline{\Psi}_{11} = 0,001374$, $\overline{\Psi}_{22} = 0,00359$, and $\overline{\Psi}_{12} = 0$. The general observation is that there is good agreement between the analytical and the finite element predictions. The nonvanishing values of $\overline{\Psi}_{ij} \equiv \overline{\Psi}^I_{ij}$ verify the existence of this constant terms. The fact that the analytical and the numerical results of the angular functions $h^I_{\alpha\beta}$ agree very well verifies the assumed symmetry conditions for $\Psi_{\alpha\beta}$ (see Section 4.8 "Symmetry conditions" in Part I).

It is worth remarking that even though the asymptotic solutions of $\Sigma_{\alpha\beta}$ have the same form as the ones of classical elasticity, the corresponding values the stress

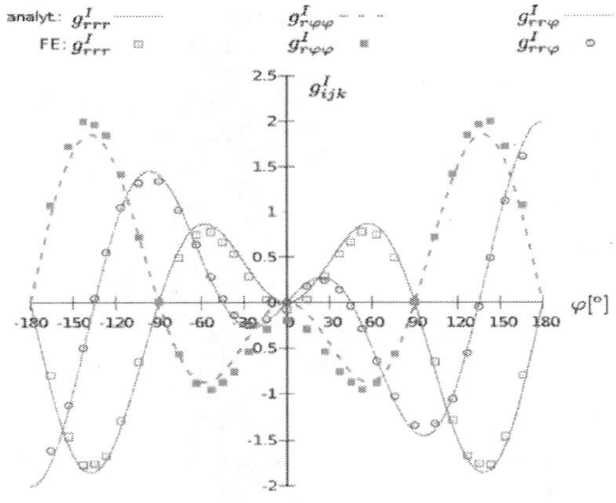

Figure 10.
Graphs of the angular functions $g^I_{r\alpha\beta}(\varphi)$ of mode-I for $\varphi \in [-180°, 180°]$.

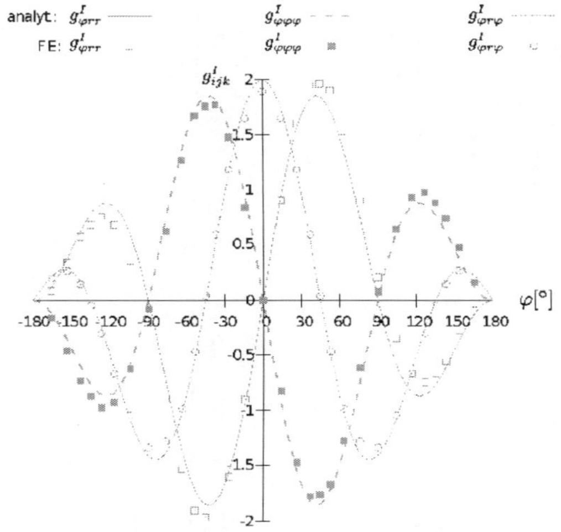

Figure 11.
Graphs of the angular functions $g^I_{\varphi\alpha\beta}(\varphi)$ of mode-I for $\varphi \in [-180°, 180°]$.

intensity factors $\tilde{K}_I$ and $\tilde{K}_{II}$ of the 3-PG Model will be, in general, different from the stress intensity factors K_I and K_{II} of classical elasticity. This difference results from the fact that the 3-PG Model and the classical elasticity imply, in general, different distributions of the components $\Sigma_{\alpha\beta}$ for identical classical boundary conditions.

Recall also, that the 3-PG Model includes two material parameters more than the classical elasticity, namely, c_1 and c_3. The effect of these nonclassical material parameters is illustrated in **Figures 12** and **13**. These figures reveal that for $c_3 =$ const. and very large values of c_1 or for $c_1 =$ const. and very small values of c_3, the values of $\tilde{K}_I$ converge to the values of K_I. In other words, the responses of the 3-PG Model approach the ones according to classical elasticity.

The effect of c_1 and c_3 on the nonclassical stress intensity factor $\tilde{L}_I$ is illustrated in **Figures 14** and **15**. The principal observation is that for smaller values of c_3, the value of $\tilde{L}_I$ gets smaller as well. On the other hand, if $c_3 =$ const., then the values of $\tilde{L}_I$ increase with increasing values of c_1.

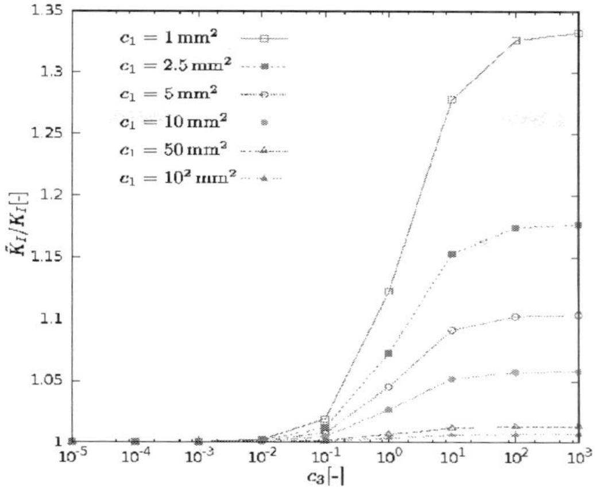

Figure 12.
Effect of the material parameter c_3 on the stress intensity factor $\tilde{K}_I$ of mode-I.

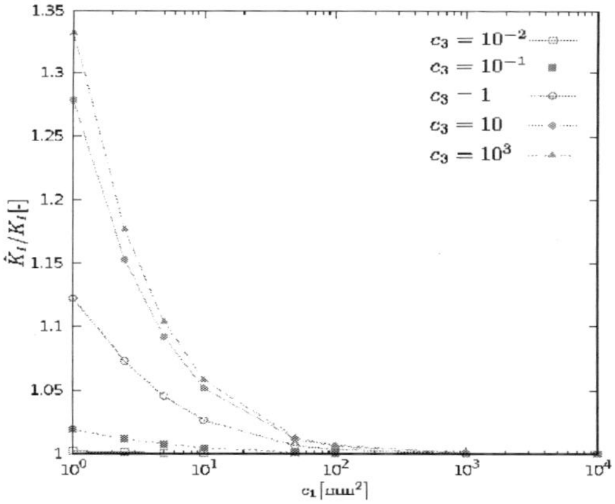

Figure 13.
Effect of the material parameter c_1 on the stress intensity factor $\tilde{K}_I$ of mode-I.

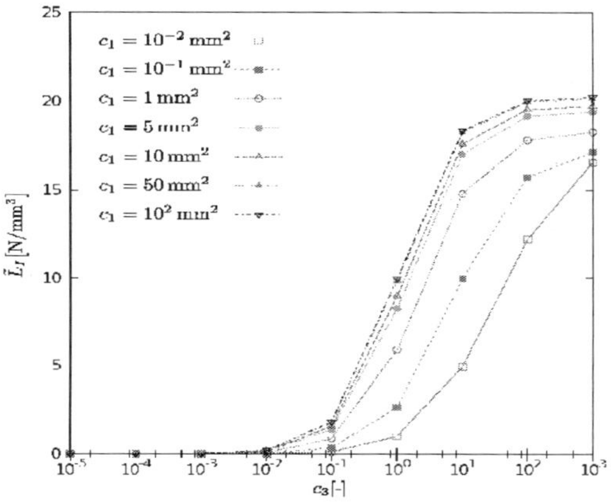

Figure 14.
Effect of the material parameter c_3 on the stress intensity factor $\tilde{L}_I$ of mode-I.

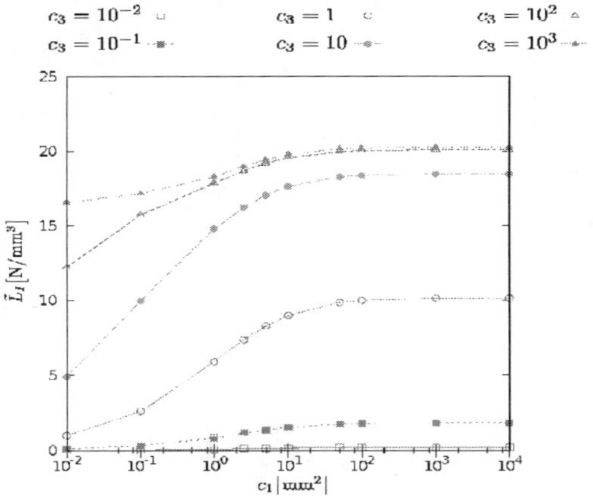

Figure 15.
Effect of the material parameter c_1 on the stress intensity factor $\tilde{L}_I$ of mode-I.

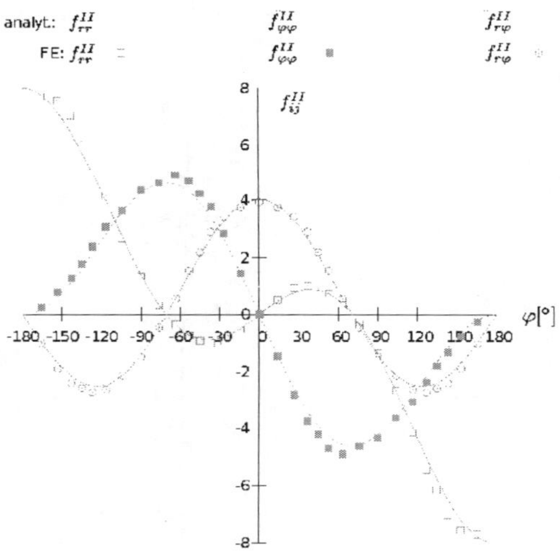

Figure 16.
Graphs of the angular functions $f^{II}_{\alpha\beta}(\varphi)$ of mode-II for $\varphi \in [-180°, 180°]$.

4.4 Results for mode-II

The graphs of the angular functions $f^{II}_{\alpha\beta}$ for both the analytical and the finite element solutions are shown in **Figure 16**. It can be seen that the graphs fit very well.

Since there are two stress intensity factors of Mode-II, $\tilde{L}_{II,1}$, and $\tilde{L}_{II,2}$, we find it convenient, in the vicinity of the crack tip, to consider the angular distributions of $\Psi_{\alpha\beta} - \overline{\Psi}_{\alpha\beta}$ and $\mu_{\alpha\beta\gamma}$ themselves instead of the corresponding distributions of the angular functions $h^{II,1}_{\alpha\beta}$, $h^{II,2}_{\alpha\beta}$, $g^{II,1}_{\alpha\beta\gamma}$, and $g^{II,2}_{\alpha\beta\gamma}$. For $r = \text{const.}$, **Figures 17–19** illustrate angular distributions of $\Psi_{\alpha\beta}$ and $\mu_{\alpha\beta\gamma}$ for both the analytical and the numerical solutions. Once more, we can recognize, that the analytical and the numerical results agree with great accuracy, which also verifies the assumed symmetry

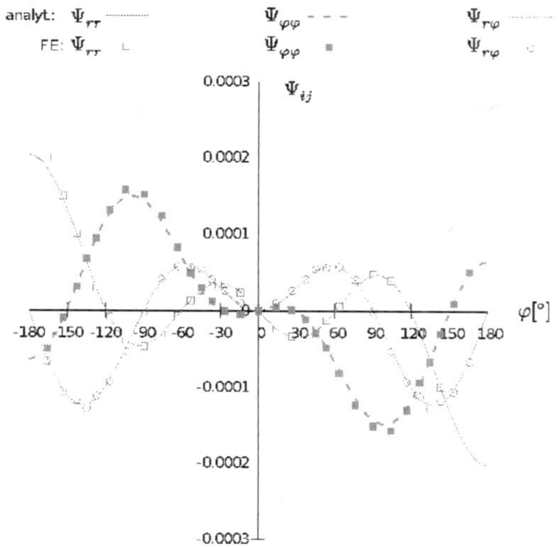

Figure 17.
Graphs of the angular distributions of $\Psi_{\alpha\beta}$ *of mode-II for a sufficiently small radius* $r = \text{const.}$ *(in vicinity of the crack tip).*

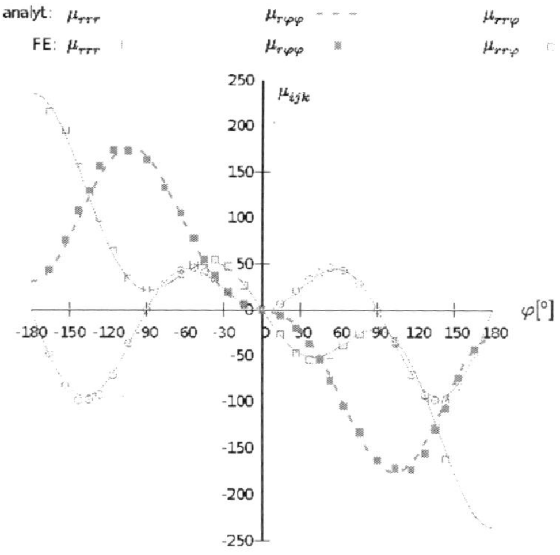

Figure 18.
Graphs of the angular distributions of $\mu_{r\alpha\beta}$ *of mode-II for a sufficiently small radius* $r = \text{const.}$ *(in vicinity of the crack tip).*

conditions of Mode-II for $\Psi_{\alpha\beta}$ (see Section 4.8 "Symmetry conditions" in Part I). Further, the values of $\overline{\Psi}_{ij}$ have been determined to be $\overline{\Psi}_{11} = 0$, $\overline{\Psi}_{22} = 0$, and $\overline{\Psi}_{12} = 0,00391$ and this, in turn, verifies the existence of the constant terms $\overline{\Psi}_{ij} \equiv \overline{\Psi}^{II}_{ij}$.

The effect of c_1 and c_3 on $\tilde{K}_{II}$ is illustrated in **Figures 20** and **21** and is similar to the effect of c_1 and c_3 on $\tilde{K}_I$ (cf. **Figures 12** and **13**).

The effect of c_1 and c_3 on the nonclassical stress intensity factors $\tilde{L}_{II,1}$ and $\tilde{L}_{II,2}$ is illustrated in **Figures 22–25**. Again, this effect is quite similar to the effect of c_1 and c_3 on $\tilde{L}_I$, but the stress intensity factors $\tilde{L}_{II,1}$ and $\tilde{L}_{II,2}$ both are negative in contrast to the positive stress intensity factor $\tilde{L}_I$.

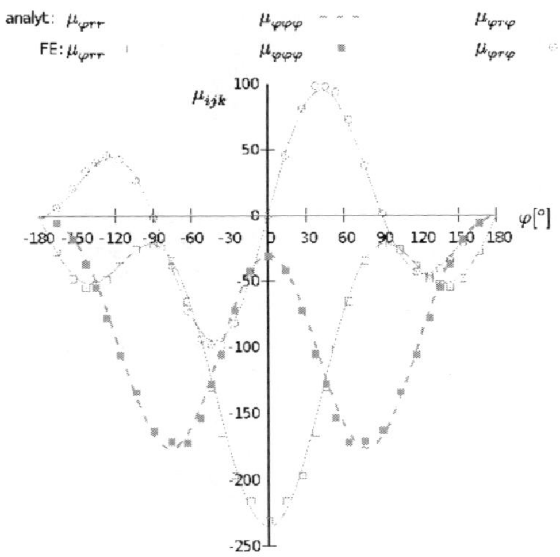

Figure 19.
Graphs of the angular distributions of $\mu_{\varphi\alpha\beta}$ of mode-II for a sufficiently small radius $r =$ const. *(in vicinity of the crack tip).*

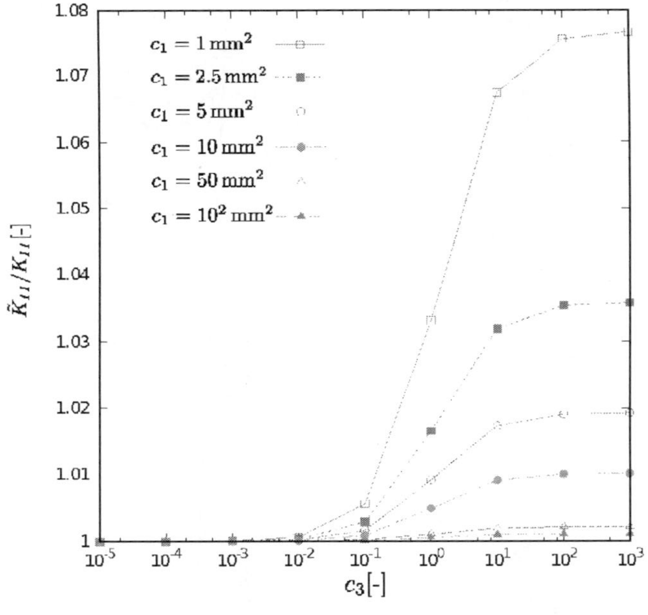

Figure 20.
Effect of the material parameter c_3 on the stress intensity factor $\tilde{K}_{II}$ of mode-II.

With regard to Mode-II crack problems, it is of interest to recall that the general solutions of $\Psi^{(0)}_{\alpha\beta}$ depend on three constants of integration, that is, $B^{(0)}$, $E^{(0)}$, and $F^{(0)}$. By virtue of the boundary conditions on the crack faces, we obtained the relation $-F^{(0)} = E^{(0)}$, which is independent of the crack geometry, that is, the crack length. We do not know further conditions to relate $B^{(0)}$ and $E^{(0)}$ with each other. However,

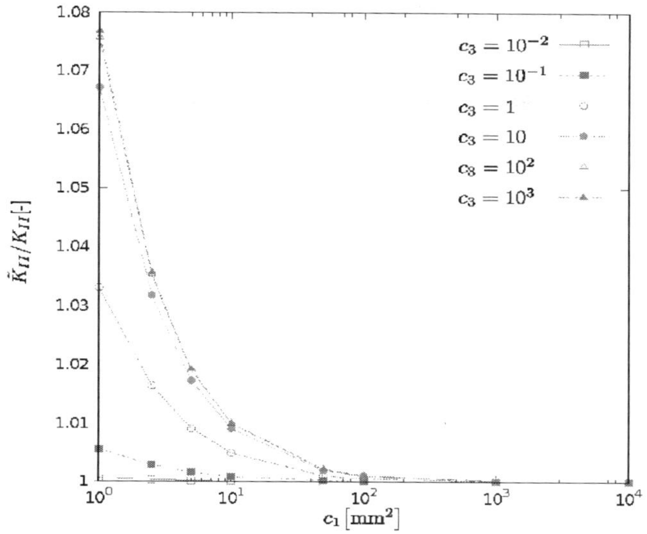

Figure 21.
Effect of the material parameter c_1 on the stress intensity factor $\tilde{K}_{II}$ of mode-II.

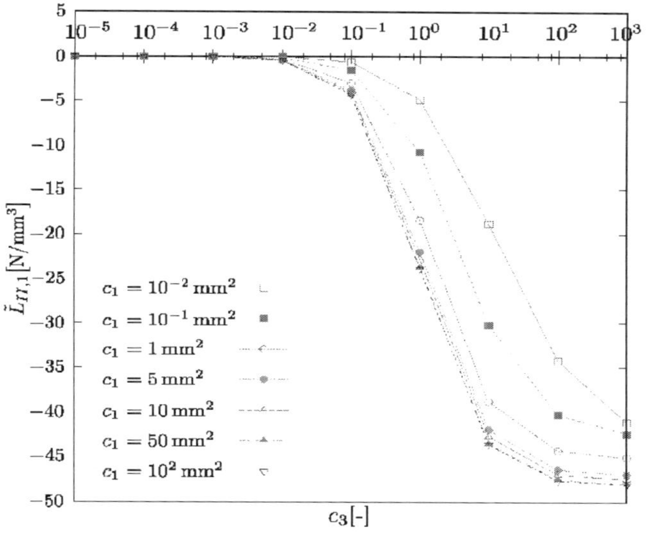

Figure 22.
Effect of the material parameter c_3 on the stress intensity factor $\tilde{L}_{II,1}$ of mode-II.

keeping in mind that, in general, stress intensity factors depend on the crack geometry and the applied loading conditions, one may ask if there exists a relation between $B^{(0)}$ and $E^{(0)}$, or equivalently between $\tilde{L}_{II,1}$ and $\tilde{L}_{II,2}$, which is independent of the crack length. To clarify this question, we consider the effect of the crack length a and the applied shear stress $\Sigma^*_{r\varphi}$ on the ratio $\frac{\tilde{L}_{II,1}}{\tilde{L}_{II,2}}$. Evidently, this ratio will depend on the material parameters c_1 and c_3. The effect of a and $\Sigma^*_{r\varphi}$ is illustrated in **Figure 26a–d**. In all figures, the width b and the length $2h$ of the specimen are the

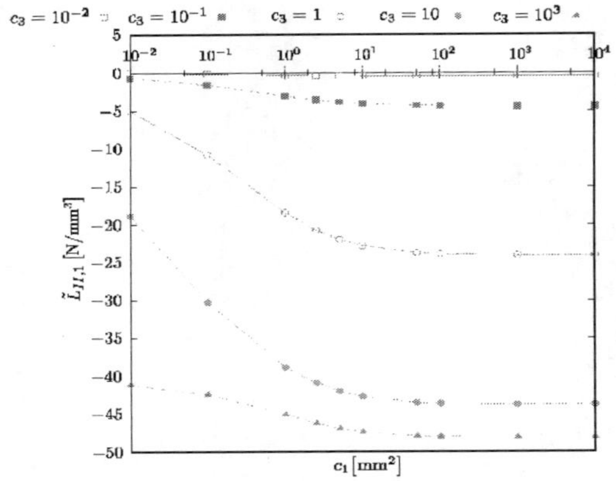

Figure 23.
Effect of the material parameter c_1 on the stress intensity factor $\tilde{L}_{II,1}$ of mode-II.

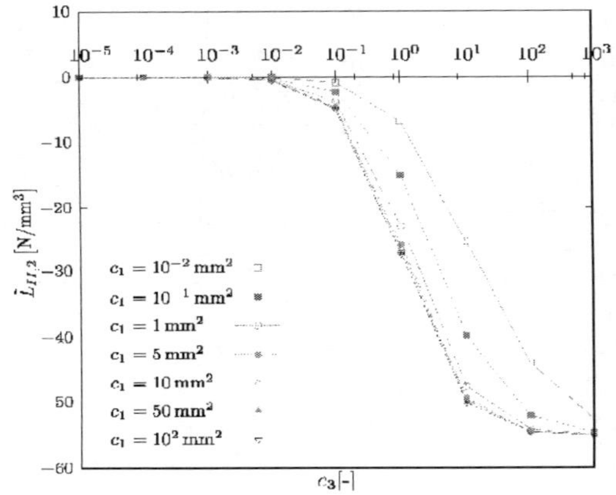

Figure 24.
Effect of the material parameter c_3 on the stress intensity factor $\tilde{L}_{II,2}$ of mode-II.

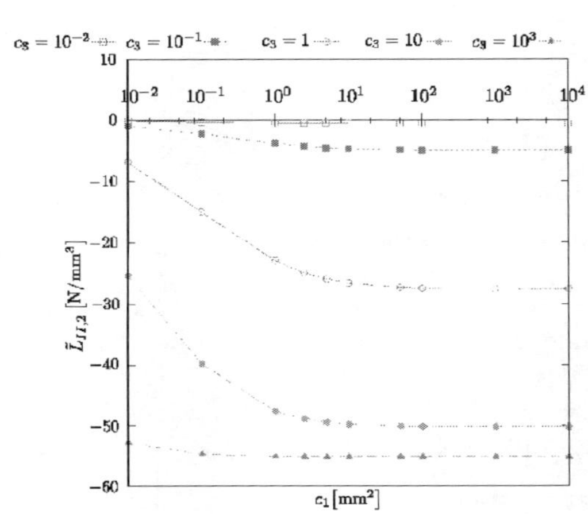

Figure 25.
Effect of the material parameter c_1 on the stress intensity factor $\tilde{L}_{II,2}$ of mode-II.

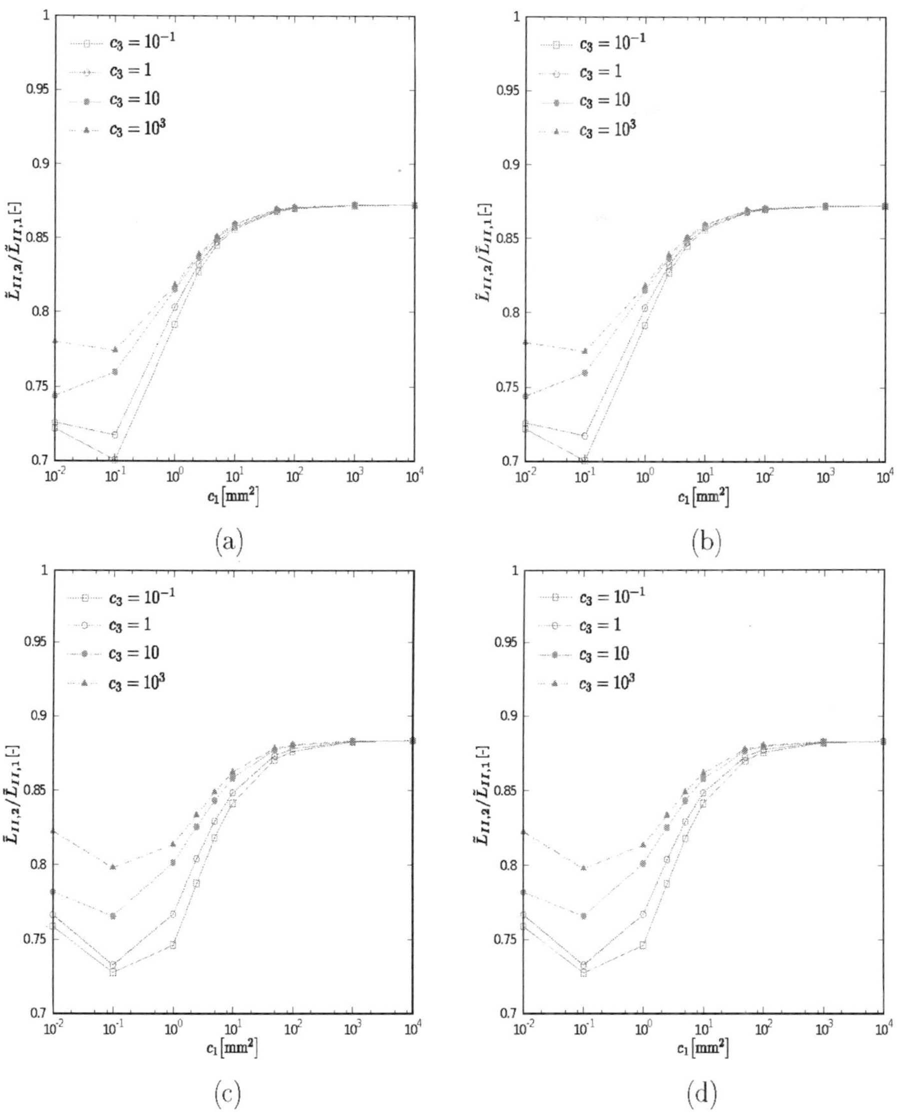

Figure 26.

Effect of the crack length a and the applied shear stress $\Sigma^*_{r\varphi}$ *on the ratio* $\frac{\tilde{L}_{II,2}}{\tilde{L}_{II,1}}$ *for mode-II crack problems.* *(a)* $a = 1$ mm, $\Sigma^*_{r\varphi} = 100$ GPa. *(b)* $a = 1$ mm, $\Sigma^*_{r\varphi} = 200$ GPa. *(c)* $a = 2$ mm, $\Sigma^*_{r\varphi} = 100$ GPa. *(d)* $a = 2$ mm, $\Sigma^*_{r\varphi} = 200$ GPa.

same and equal to the values given in Section 4.1. However, a and $\Sigma^*_{r\varphi}$ are different for each figure. It can be recognized from **Figure 26a** and **b** that the ratio $\frac{\tilde{L}_{II,1}}{\tilde{L}_{II,2}}$ does not depend on the applied loading, $\Sigma^*_{r\varphi}$, which might be expected, for the problem is linear. But the ratio $\frac{\tilde{L}_{II,1}}{\tilde{L}_{II,2}}$ depends on the crack length, a, as can be seen by comparing **Figure 26a** and **c** or **Figure 26b** and **d**. This result may be seen as a justification of considering $\tilde{L}_{II,1}$ and $\tilde{L}_{II,2}$ as different stress intensity factors.

5. Concluding remarks

A finite element model for the 3-PG Model has been developed and verified with reference to the expected responses of a square section with a hole subjected to displacement-controlled tension loading. Then, the finite element model has been employed to discuss Mode-I and Mode-II crack problems. From these investigations, one can draw the following conclusions.

1. The finite element and the analytical solutions fit with good accuracy.

2. In particular, the finite element simulations confirm the order of singularity, $r^{-\frac{1}{2}}$, and the assumed symmetry conditions for the micro-deformation very well. Further, they confirm the existence of a constant term in the asymptotic expansion of the micro-deformation.

3. By using numerical simulations with the developed finite element model, the effect of the nonclassical material parameters on the classical and the nonclassical stress intensity factors has been investigated. Especially, limiting cases of material parameters leading to responses of classical elasticity have been analyzed.

4. For Mode-II loading conditions, two independent stress intensity factors have been assumed to be present in the near-tip fields of the double stress. The numerical investigation seem to verify this remarkable feature.

Abbreviations

eq. equation

References

[1] Williams ML. On the stress distribution at the base of a stationary crack. Journal of Applied Mechanics. 1957;**24**:109-114

[2] Diegele E, Elsässer R, Tsakmakis C. Linear micropolar elastic crack-tip fields under mixed mode loading conditions. International Journal of Fracture. 2004; **129**:309-339

[3] Hughes TJR. The Finite Element Method: Linear Static and Dynamic Finite Element Analysis. Englewood Cliffs, New Jersey: Prentice-Hall; 1987

[4] Gould GL. Introduction to Linear Elasticity. 2nd ed. New York: Springer-Verlag; 1994

[5] Henshell RD, Shaw KG. Crack tip finite elements are unnecessary. International Journal for Numerical Methods in Engineering. 1975;**9**:495-507

[6] Barsoum RS. On the use of isoparametric finite elements in linear fracture mechanics. International Journal for Numerical Methods in Engineering. 1976;**10**:25-37

Chapter 5

Synthesis of WO_3 Nanostructures and Their Nanocomposites with Graphene Derivatives via Novel Chemical Approach

Abstract

Trioxide Tungsten (WO_3), an n-type semiconductor that exhibits a wide band gap of 2.5 to 3.6 eV, has attracted special attention from the scientific community. This attraction is due to its manifold properties, which not only follow the development of technologies, but accelerate it. There are several methods to synthesize WO_3 nanostructures with various morphologies. In the present study, for the first time, a novel chemical method was developed for the preparation of WO_3 nanostructures by using tungsten carbide (WC) as precursor. This novel approach has many advantages such as high yields, simple methodology and easy work up. Moreover, graphene oxide coated with WO_3 nanostructured is prepared via in-situ and ex-situ chemical approaches followed by subsequent thermal treatment at 500°C. The obtained samples were characterized by different techniques to confirm the transformation of WC to WO_3 nanostructures and the formation of their nanocomposites with graphene derivatives.

Keywords: trioxide tungsten, tungsten carbide, nanocomposites, graphene oxide

1. Introduction

Nanocomposites based on nanocarbons and nanostructured metal oxides (NMO) offer the possibility of improving the performance of several devices and developing multifunctional systems by combining the properties of each individual phase. The importance of choosing an appropriate route for the preparation of these nanocomposites has led scientists to take an interest in the development of synthetic methods that are versatile, generalized and easily adaptable to prepare various nanocomposites [1].

Among various NMO, nano-sized (WO_3) structures are of great interest due to their stability in aqueous solution, good crystal quality, remarkable charge transport and unique optical properties [2]. With these manifold properties, they have already been widely used in photocatalysts, photonics and electronics devices [3]. For most of these applications, many physical and chemical approaches have been developed to synthesize WO_3 nanostructures using common precursors such as Na_2WO_4 and WCl_6 [4]. JAMALI et al. recently have reported sol-gel method and its effect on the structural and morphological properties of WO_3 nanostructures [5].

As well as, Tehrani et al. have prepared 1D and 2 D WO_3 by hydrothermal treatment [6]. However, to the best of our knowledge this is the first study about the synthesis of WO_3 nanostructures using tungsten carbide WC as precursor through chemical oxidation approach.

The combination of these nanostructures with nanocarbons in general and graphene derivatives in particular have attracted considerable interest in the scientific community. Graphene oxide (GO) is a single or few layers of oxidized graphite, just like graphene considered as a single layer of graphite [7]. It has been defined as an important precursor of graphene and the basic material for the development of graphene-based nanocomposites [8]. Furthermore, the synthesis of these kinds of nanocomposites having a well-defined structure represents a problem with the methods requiring particular conditions. The key challenge, to improve their performance and broaden their field of application, is to develop simpler synthetic methods in order to increase the control during the formation and the anchoring of nanostructures on the surface of nanocarbons, while maintaining the structural integrity of the composite at the nanoscale.

To date, nanocomposites with different morphologies can be prepared by various methods that can be classified into two categories: ex situ hybridization and in situ crystallization [9]. Ex-situ hybridization involves mixing the pre-synthesized nanocarbons and NMO [10]. Prior to mixing, surface modification of nanostructures and/or nanocarbons is often required, so that they can bind either through covalent or non-covalent interactions including Van der Waals interactions, hydrogen bonds or electrostatic interactions. In addition, the type of functionalization and the strength of the interaction determine the distribution of metal oxide nanoparticles on the surface of the nanocarbons. Although ex-situ hybridization is able to pre-select nanostructures with desired functionality, it sometimes suffers from the low density and non-uniform distribution of nanostructures on nanocarbon surfaces. However, in-situ crystallization can give rise to uniform surface distribution of nanoparticles by controlling nucleation sites on nanocarbons via surface functionalization. As a result, a continuous film of nanoparticles on surfaces can be achieved [11].

In this work, we developed for the first time, a new and simple method to synthesize the nanostructures of WO_3 and their nanocomposites by in situ and ex situ chemical approaches. The synthesis process is based on simple chemical oxidation and subsequent thermal annealing. Moreover, in order to validate the feasibility of our approaches, we examine our samples obtained by different techniques such as X-ray diffraction (XRD) to analyze structural properties, scanning electron microscopy (SEM) to determine morphology, shape and size of nanostructures, as well as Fourier transform infrared (FTIR) to elucidate the vibrational behavior and the type of interaction between the different constituents.

2. WO_3 nano-sized structures

2.1 Synthesis method

The synthesis of WO_3 nanostructures involved two steps. In the first step, 1 g of WC and 10 ml of H_2SO_4 were mixed and stirred in an ice bath. This mixture was continuously stirred while 3 g of $KMnO_4$ was added slowly over 1 h, the temperature was kept below 20°C. After adding $KMnO_4$, the mixture was stirred for a further 2 h, 15 ml H_2O_2 (30 wt% aqueous solution) was slowly added and then the resulting solution turned into homogenous yellow color and was left to stir for another 2 h. The product was centrifuged and dried in the oven at 60°C. In the second step, the obtained yellow powder corresponding to tungsten oxide hydrate $WO_3.H_2O$ was dispersed in a mixture of water and ethanol (1, 5), dropped onto a glass substrate and heated in a furnace up to 500°C at the heating rate of $10°cm^{-1}$. The sample was kept inside the furnace at the mentioned temperature for 5 h [12].

2.2 Characterization

Structural information of different tungsten-based materials are shown in **Figure 1**. The WC diffraction spectrum shows three major intense peaks located at 2θ = 31.57°, 35.72° and 48.40° which correspond well to the crystallographic planes (001, 100, 101) of WC, respectively. The positions of the peaks are in good agreement with the JCPDS reference (N: 65–4539), which confirm the hexagonal structure of WC, of space group P6m2, with the parameters a = 2.906 Å and c = 2.838 Å. According to a study carried out on the oxidation of WC to WO_3 via a dry synthetic route in air and at high temperature [13], the authors found that tungsten carbide oxidizes more quickly than tungsten metal. However, the oxidation WC chemical with precise size control has never been reported to our knowledge.

After oxidation, we can identify the crystalline nature of the prepared sample by analyzing its X-ray diffraction (XRD) spectrum. All the diffraction peaks are well indexed to tungsten oxide monohydrate $WO_3.H_2O$ with an orthorhombic phase (JCPDS N: 43–0679, of space group: Pmnb (62) having the lattice parameters: a = 5.238 Å, b = 10.704 Å and c = 5.12 Å). The peaks are intense and narrow, indicating better crystallinity. Additionally, no peak of impurities and/or other hydrated phase such as $WO_3.2H_2O$ and $WO_3.0.33H_2O$ was observed in the spectrum [14, 15]. Indeed, this material has been widely used as a precursor for the synthesis of the monoclinic and hexagonal phases of WO_3, which are different in their geometry and their stability [16]. The hexagonal phase is metastable, while the monoclinic phase is the most stable than all other structures of WO_3 due to its pseudo-cubic structure made up of a three-dimensional array of WO_6 octahedra [17].

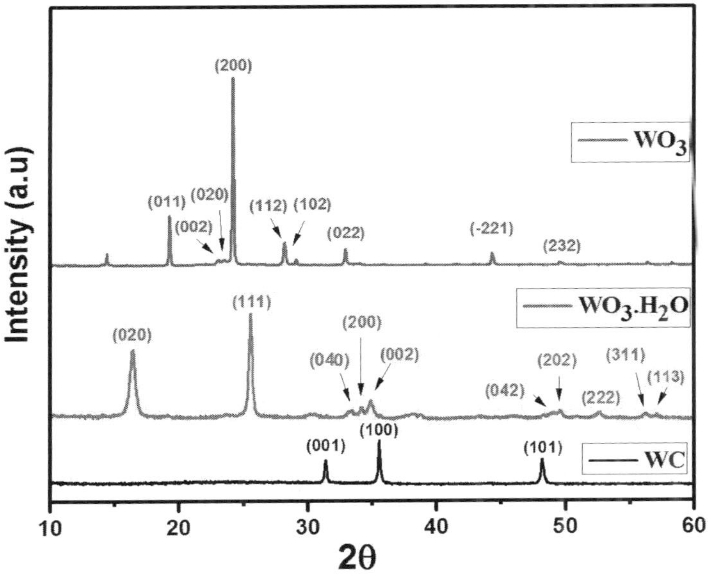

Figure 1.
X-ray diffraction patterns of WC, $WO_3.H_2O$, WO_3 nanostructures.

During the heat treatment at 500°C, $WO_3.H_2O$ as synthesized gradually dehydrates until it turns completely into WO_3. **Figure 1** blue spectrum shows a series of diffraction peaks at 2θ = 19.26°, 23.07°, 23.45°, 24.17°, 28.22°, 29.13°, 32.86°, 44.3° and 49.6° which correspond to the crystallographic planes (011, 002, 020, 200, 112, 102, 022, 320, 232). The positions of these peaks are consistent with those expected for the WO_3 monoclinic phase reported in the JCPDS N: 43–1035 file, of space group: P21/n (14) and lattice parameters: a = 7.297 Å, b = 7.539 Å and c = 7.688 Å.

The crystallite size was calculated, using the Scherrer equation, for all tungsten-based components, as shown in **Table 1**. The results obtained demonstrate the growth in the size of the crystallites of WO_3 as a function of the annealing temperature. Gui et al. synthesized WO_3 with monoclinic phase using the hydrothermal route and WCl_6 as a precursor, they estimated a particle size of the order of 22 nm [18]. Fu et al. reported a value, similar to our result, of around 42 nm hydrothermally treated at 180°C for 24 h, using $Na_2WO_4.2H_2O$ as a precursor [19].

Material	Plan	2θ	Intensity	Size (nm)	Average size (nm)
WC	(001)	31.4°	0.56	39.32	36
	(100)	35.4°	0.98	39.28	
	(101)	48.2°	0.57	28.4	
$WO_3.H_2O$	(020)	16.4°	0.66	17.28	21
	(111)	25.4°	0.98	31.78	
	(040)	33.2°	0.11	18.07	
	(200)	34.1°	0.14	36.18	
	(002)	34.9°	0.19	17.65	
WO_3 (500°C)	(011)	13.31°	0.27	50.2	57
	(002)	23°	0.042	62	
	(020)	23.4°	0.038	67	
	(200)	24.1°	0.98	56.18	
	(112)	28.19°	0.12	50.65	

Table 1.
Crystallites size of tungsten-based nanomaterials.

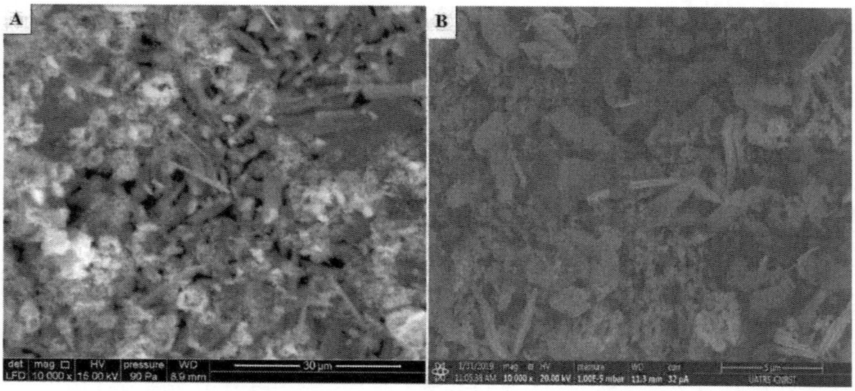

Figure 2.
SEM images of WO_3 nanostructures.

The morphology observation is performed to investigate the surface and the shape of the nanostructured WO_3. As seen from the SEM image (**Figure 2A**). The morphology obtained after a heat treatment at 500°C is effectively homogeneous and composed of what are called nanorods, with the presence of a few hollow nanospheres. In **Figure 2B**, WO_3 nanorods are clearly observed. Choi et al. demonstrated different morphologies of WO_3 such as nanowires, nanorods and nanosheets by varying the volume of water (%) and ethanol [20]. Marques et al. have prepared different morphologies including nanoplates and nanoflowers by varying the initial precursor and the pH of the medium [21].

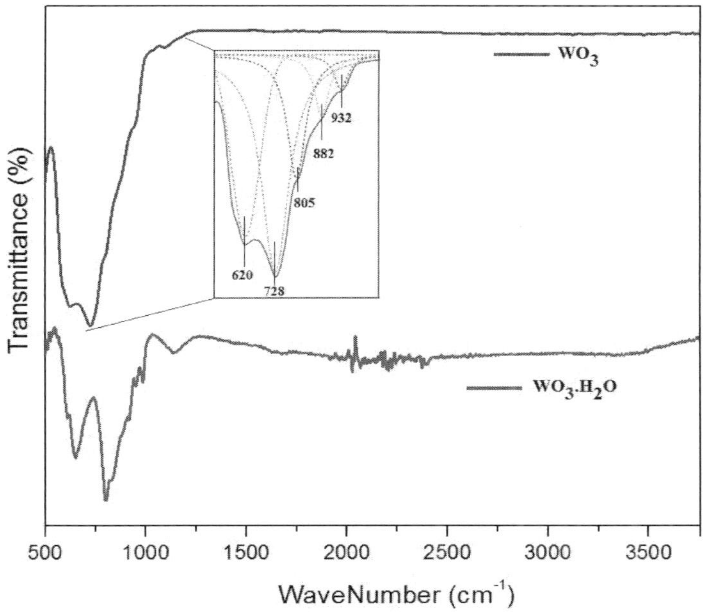

Figure 3.
FT-IR spectra of $WO_3.H_2O$, WO_3 nanostuctures.

Material	Frequency (cm^{-1})	Attribuation
$WO_3.H_2O$	602	γ (W-O-W)
	650	δ (O-W-O)
	883	ν (W-O-W)
	985	ν (W=O)
	1139	ν (W-OH)
	3440	ν (-OH)
WO_3 (500 °C)	620	γ (W-O-W)
	728	δ (O-W-O)
	805	ν (O-W-O)
	882	ν (W-O-W)
	932	ν (W=O)

Table 2.
Tungsten oxide infrared bands and their attributions.

In order to determine the exact composition of the sample, and to gain more information on its crystal structure, FTIR measurement was implemented.

In the range between 500 and 1000 cm^{-1} (**Figure 3**), multi-bands are observed, attributed to tungsten-oxygen elongation vibrations including elongation vibrations (ν), in-plane bending vibrations (δ) and out of plane (γ), as shown in **Table 2**. These characteristic bands prove the successful synthesis of WO_3 monoclinic phase.

3. GO/WO_3 nanocomposites

3.1 GO/WO_3 prepared in-situ

3.1.1 Synthesis method

For the in situ method, the nanocomposite was prepared using the same process as that used for WO_3; but with the presence of graphite powder. As shown in **Figure 4**, after the interaction of WC with $KMnO_4$ and H_2SO_4 solution for 1 h in an ice bath, a small amount of graphite powder was added and the mixture kept under mechanic agitation for 2 h. Then, H_2O_2 was added and the resulting solution turned into green-brown color. The homogeneous solutions obtained were slowly dropped onto glass substrate, and heated at 500°C for 5 h.

3.1.2 Characterization

The obtained XRD patterns are shown in **Figure 5**. In the case of the nanocomposite prepared in situ from graphite and WC, we note that the position and intensity of the peaks obtained have been greatly modified. The main peaks of the monoclinic phase are reduced in intensity and several peaks have appeared corresponding to other polymorphs of WO_3, including the hexagonal, orthorhombic phases and substoichiometric chemical compositions [22, 23].

The appearance of a similar multi-phase structure is consistent with the fact that the in situ insertion of graphite leads to the modification of the crystal structure formed of tungsten oxide. Some of the more well-known non-stoichiometric compositions of WO_x are $W_{17}O_{48}$, $W_{20}O_{58}$, $W_{18}O_{49}$ and $W_{24}O_{68}$. Such oxides are formed by the sharing of vertices of WO_6 octahedra, which alternate with those which are partially established by sharing the edges [24].

Oxygen removal occurs through the crystal shear mechanism and as the x value in WOx decreases, the WO_6 octahedra groups form closer shear planes. However, for x values close to 3, these planes are considered as extended defects. On the other hand, with a further decrease in x, the shear planes tend to interact with each other and align in parallel, filling the space between them with sharing vertices of WO_6 octahedra. For x less than or equal to 2.87, the structure becomes unstable

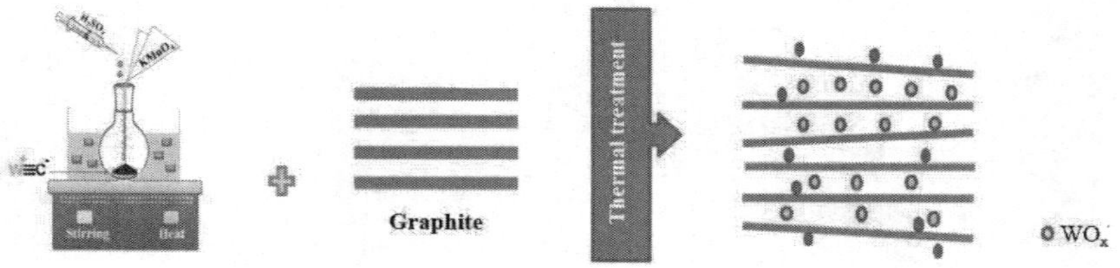

Figure 4.
Preparation method of in-situ nanocomposite.

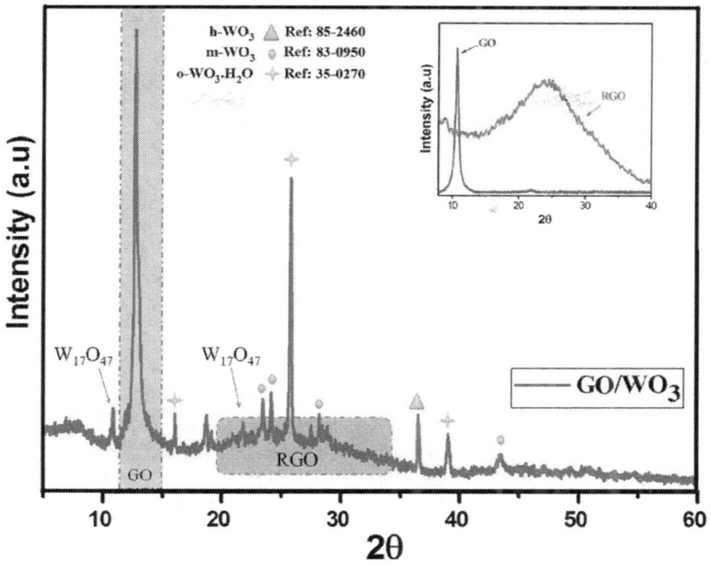

Figure 5.
X-ray diffraction of GO/WO$_3$ in-situ prepared nanocomposite.

and further restructuring occurs involving the formation of pentagonal columns parallel to the monoclinic axis that are either single or paired by partitioning of the edges. Moreover, recent studies [25] have shown that non-stoichiometric tungsten oxide nanostructures have better performance in advanced applications. They have shown that the oxygen deficiencies presented in these nanomaterials can be major assets for storing energy. In addition, the X-ray diffraction pattern of the nanocomposite shows an intense peak characteristic of graphene oxide located at 2θ = 12.73° and assigned to the (002) plane with an interlayer "d" spacing of 6.95 Å, which is quite high compared to graphite (3.36 Å) and slightly lower than that of GO (7.93 Å) obtained by the Hummer method and in the absence of WC. The appearance of a small and wide band around 23–26° confirms the formation of a few partially oxidized graphite nanosheets.

Indeed, the modification of the spacing between the layers of graphite is directly correlated to their exfoliation through the introduction of the metal oxide nanostructures formed which also causes the modification of the Van der Walls interaction. In addition, it should be noted that the appearance of a multiphase structure could be due to the fact that the in-situ insertion of graphite has a strong influence on the orientation and length of the tungsten-oxygen bond, resulting in non-stoichiometric WOx nanostructures.

SEM image is used to investigate the microscopic structure of the obtained sample. **Figure 6** shows the morphological aspect of in situ GO/WO$_3$ nanocomposite. The presence of graphene layers functionalized by spherical-shaped WO$_3$ nanostructures distributed randomly can be detected. This nanocomposite has a stable structure reducing the agglomeration of graphitic planes and indicating the interfacial interaction between the two components. These results are in good agreement with those obtained by X-ray diffraction.

The infrared spectrum of nanocomposite prepared in situ (**Figure** 7) exhibits a broad absorption band at 640 cm^{-1} attributed to the W-O-W vibration, and another band at 982 cm^{-1} characteristic of the short elongation bond W = O [26]. This is explained by the fact that the in situ preparation of tungsten oxide nanostructures

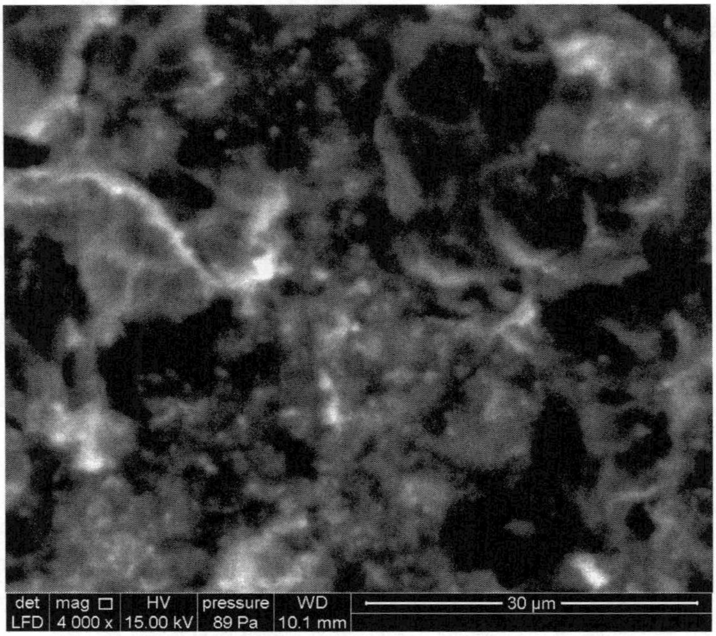

Figure 6.
SEM image of GO/WO_3 in-situ prepared nanocomposite.

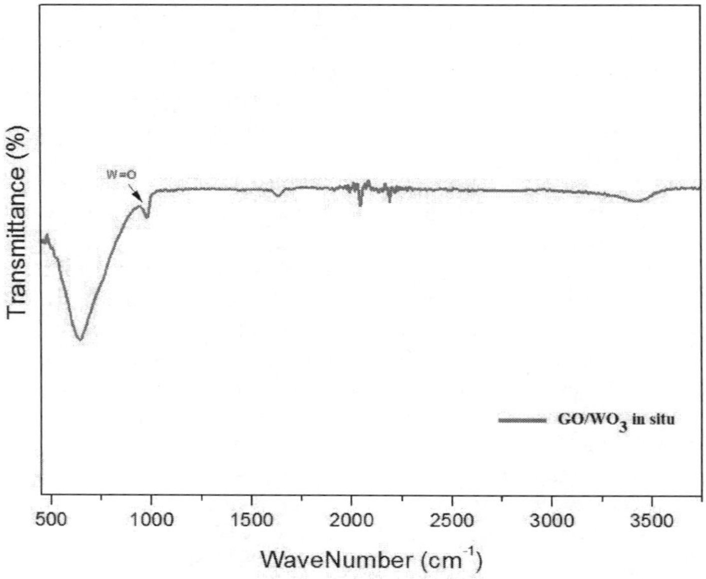

Figure 7.
FT-IR spectrum of GO/WO_3 in-situ prepared nanocomposite.

in the presence of graphite leads to the formation of non-stoichiometric WOx nanostructures. In fact, during the interaction, the vibration modes linked to the single WO bonds gradually weakened, while the W = O double bonds are formed by increasing the areas of the boundaries and creating more deficit in atoms of oxygen, which contributes to the modification of the crystallographic structure. These results are in good agreement with those obtained by XRD.

3.2 GO/WO_3 prepared ex-situ

3.2.1 Synthesis method

GO/WO_3 nanocomposite was synthesized by mixing solutions of hydrated tungsten oxide and graphene oxide. In a typical procedure, the solution of WO_3. H2O was mixed with that of GO, then the resulting mixture is placed in an ultrasonic bath for 30 min at room temperature. After deposition on the glass substrate and calcination at 500°C., the GO/WO_3 nanocomposite was obtained with a mass ratio of 1: 1 (**Figure 8**).

3.2.2 Characterization

The XRD spectrum in **Figure 9** reveals the crystal structure of the GO/WO_3 nanocomposites prepared ex situ and treated at 500°C. The diffractogram of the nanocomposite is dominated by the peaks characteristic of the monoclinic phase of WO_3 [JCPDS N: 26–0575]. The three main peaks at 2θ = 23.07°, 23.45°, 24.17° characteristic of this phase are overlapped with the broad band corresponding to reduced GO.

The same behavior was obtained for the RGO/WO_3 nanocomposite prepared by the hydrothermal method, as well as GO/ZnO and GO/TiO_2 nanocomposites prepared ex situ [22–24]. It is well known that the formation of reduced GO is confirmed by the disappearance of the intense peak in GO at 10.9° and the appearance of a larger one around 26°. Due to the oxygenated functional groups on the surface of the carbon planes, aromatic regions with sp^3 networks provide active sites for interacting with other chemical species through interactions at interfaces. Thus, GO is a very important precursor for the preparation of graphene-based composite materials with metals, metal oxides, polymers and CNTs for various applications. Consequently, this result obtained confirms the reduction of GO after the functionalization with WO_3 and the formation of the nanocomposite. Furthermore, no secondary phase was detected, suggesting that the nanostructured WO_3 retained its monoclinic phase after its interaction with the GO nanosheets.

For the morphology of the GO/WO_3 nanocomposite prepared ex situ (**Figure 10**), we observe the presence of the one-dimensional shape with a homogeneous and uniform distribution. The nanorods formed in the presence of GO have lengths lower than those of the WO_3 nanorods alone. In addition, the trace of the GO sheets is not clearly observed, confirming that the GO sheets are covered by the nanostructures of WO_3.

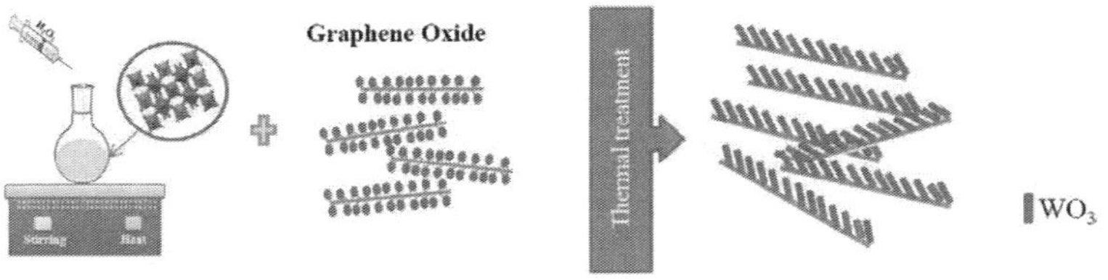

Figure 8.
Preparation method of ex-situ nanocomposite.

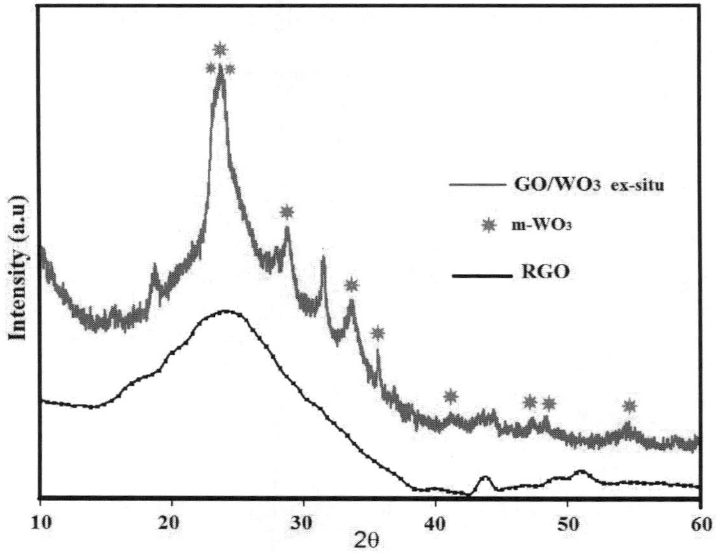

Figure 9.
X-ray diffraction of GO/WO_3 ex-situ prepared nanocomposite.

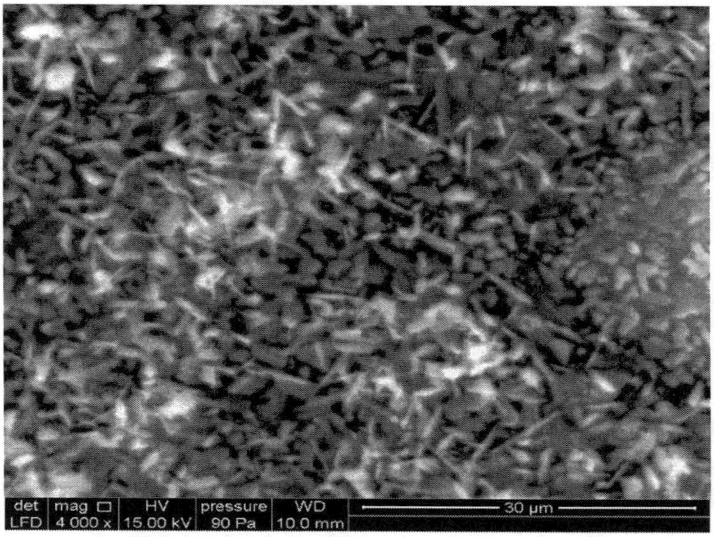

Figure 10.
SEM image of GO/WO_3 ex-situ prepared nanocomposite.

In order to determine the exact composition of the samples as well as the types of interactions between the constituents, the Fourier Transform IR Spectroscopy (FTIR) analysis was performed. In the low-frequency region (**Figure 11**), the nanocomposite prepared ex-situ exhibits the characteristic bands of the W-O bond. In addition, a new band appears at around 1120 cm^{-1} attributed to the W-O-C link [26]. This result demonstrates that WO_3 nanorods uniformly attach to the GO surface through covalent functionalization.

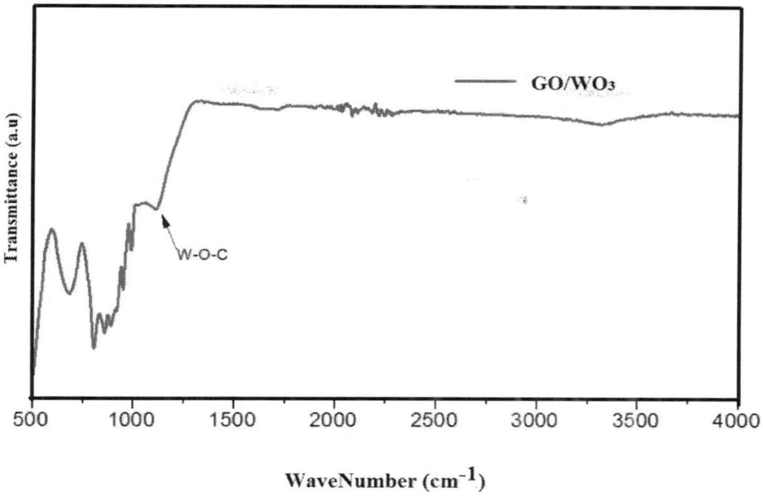

Figure 11.
FT-IR spectra of GO/WO_3 ex-situ prepared nanocomposite.

4. Conclusion

In summary, we carried out a study of structural, morphological and vibrational properties of tungsten oxide nanostructures and their nanocomposite with GO nanosheets. First of all, we succeeded in synthesizing WO_3 through a new methodology based on the oxidation of WC, using strong acids and oxidizing agents, which allowed us to obtain the hydrated tungsten oxide $WO_3.H_2O$ as an intermediate product, before being completely transformed by heat treatment at high temperature in air into WO_3. By using XRD, we were able to identify the different structures, as well as the variation in size of tungsten-based crystallites. In the case of the prepared nanocomposites, we found that graphite and GO have a significant effect on the crystallographic structure, morphology and stoichiometry of the nanostructured tungsten oxide. As result, the in situ prepared WO_3 nanostructures have shown a drastic change in the morphology and the stoichiometry when their growth is initiated in the presence of graphite powder. However, the ex situ preparation of the composite leads to the formation of well-dispersed WO_3 nanorods, with monoclinic phase, covalently bonded to the RGO nanosheets. This study provides a possibility for preparing tungsten oxide nanostructures based nanocomposites with low-cost and no special equipment.

References

[1] T. Guo, M.-S. Yao, Y.-H. Lin, C.-W. Nan, A comprehensive review on synthesis methods for transition-metal oxide nanostructures, Cryst. Eng. Comm. 17 (2015) 3551-3585.

[2] M. Kumar, Metal oxide nanostructures: growth and applications, Adv. Nanomater. 79 (2016) 203-230.

[3] W. Zhao, B. Cui, H. Qiu, P. Chen, Y. Wang, Multifunctional Fe3O4@ WO3@ mSiO2–APTES nanocarrier for targeted drug delivery and controllable release with microwave irradiation triggered by WO3, Mater. Lett. 169 (2016) 185-188.

[4] R. Lei, H. Ni, R. Chen, B. Zhang, W. Zhan, Y. Li, Hydrothermal synthesis of WO3/Fe2O3 nanosheet arrays on iron foil for photocatalytic degradation of methylene blue, J. Mater. Sci. Mater. Electron. 28 (2017) 10481-10487.

[5] W. Wu, Q. Yu, J. Lian, J. Bao, Z. Liu, S.-S. Pei, Tetragonal tungsten oxide nanobelts synthesized by chemical vapor deposition, J. Cryst. Growth 312 (2010) 3147-3150.

[6] S. Bai, K. Zhang, L. Wang, J. Sun, R. Luo, D. Li, A. Chen, Synthesis mechanism and gas-sensing application of nanosheet-assembled tungsten oxide microspheres, J. Mater. Chem. A. 2 (2014) 7927-7934.

[7] Y.M. Shirke, S.P. Mukherjee, Selective synthesis of WO3 and W18O49 nanostructures: ligand-free pH-dependent morphology-controlled self-assembly of hierarchical architectures from 1D nanostructure and sunlight-driven photocatalytic degradation, Cryst. Eng. Comm. 19 (2017) 2096-2105.

[8] A.S. Kurlov, A.I. Gusev, Oxidation of tungsten carbide powders in air, Int. J. Refract. Metals Hard Mater. 41 (2013) 300-307.

[9] L. Wang, Z. Wei, M. Mao, H. Wang, Y. Li, J. Ma, Metal oxide/graphene composite anode materials for sodium-ion batteries, Energy Stor. Mater. 16 (2019) 434-454.

[10] S.G. Chatterjee, S. Chatterjee, A.K. Ray, A.K. Chakraborty, Graphene–metal oxide nanohybrids for toxic gas sensor: a review, Sensor. Actuator. B Chem. 221 (2015) 1170-1181.

[11] R. Hatel, M. Goumri, B. Ratier, M. Baitoul, Graphene derivatives/Fe3O4/polymer nanocomposite films: optical and electrical properties, Mater. Chem. Phys. 193 (2017) 156-163.

[12] J. Kaur, K. Anand, K. Anand, R.C. Singh, WO3 nanolamellae/reduced graphene oxide nanocomposites for highly sensitive and selective acetone sensing, J. Mater. Sci. 53 (2018) 12894-12907.

[13] M. Khenfouch, U. Buttner, M. Baïtoul, M. Maaza, Synthesis and characterization of mass produced high quality few layered graphene sheets via a chemical method, Graphene 3 (2) (2014) 7-13.

[14] M. Ahmadi, M.J.-F. Guinel, Synthesis and characterization of tungstite (WO3·H2O) nanoleaves and nanoribbons, Acta Mater. 69 (2014) 203-209.

[15] D. Sandil, S. Srivastava, B.D. Malhotra, S.C. Sharma, N.K. Puri, Biofunctionalized tungsten trioxide-reduced graphene oxide nanocomposites for sensitive electrochemical immunosensing of cardiac biomarker, J. Alloy. Comp. 763 (2018) 102-110.

[16] M.A. Fraga, A. Contin, L.A.A. Rodríguez, J. Vieira, R.A. Campos, E.J. Corat, V.J.T. Airoldi, Nano-and microcrystalline diamond deposition on pretreated WC–Co substrates:

structural properties and adhesion, Mater. Res. Express 3 (2016) 25601.

[17] V. Hariharan, S. Radhakrishnan, M. Parthibavarman, R. Dhilipkumar, C. Sekar, Synthesis of polyethylene glycol (PEG) assisted tungsten oxide (WO3) nanoparticles for L-dopa bio-sensing applications, Talanta 85 (2011) 2166-2174.

[18] L. Ghasemi, H. Jafari, Morphological characterization of tungsten trioxide nanopowders synthesized by sol-gel modified pechini's method, Mater. Res. 20 (2017) 1713-1721.

[19] V. Hariharan, V. Aroulmoji, K. Prabakaran, B. Gnanavel, M. Parthibavarman, R. Sathyapriya, M. Kanagaraj, Magnetic and electrochemical behaviour of cobalt doped tungsten oxide (WO3) nanomaterials by microwave irradiation method, J. Alloy. Comp. 689 (2016) 41-47.

[20] J. Jayachandiran, A. Raja, M. Arivanandhan, R. Jayavel, D. Nedumaran, A facile synthesis of hybrid nanocomposites of reduced graphene oxide/ZnO and its surface modification characteristics for ozone sensing, J. Mater. Sci. Mater. Electron. 29 (2018) 3074-3086.

[21] M. Shanmugam, A. Alsalme, A. Alghamdi, R. Jayavel, In-situ microwave synthesis of graphene–TiO2 nanocomposites with enhanced photocatalytic properties for the degradation of organic pollutants, J. Photochem. Photobiol. B Biol. 163 (2016) 216-223.

[22] C. Chacón, M. Rodríguez-Pérez, G. Oskam, G. Rodríguez-Gattorno, Synthesis and characterization of WO3 polymorphs: monoclinic, orthorhombic and hexagonal structures, J. Mater. Sci. Mater. Electron. 26 (2015) 5526-5531.

[23] W.-H. Hu, G.-Q. Han, B. Dong, C.-G. Liu, Facile synthesis of highly dispersed WO3·H2O and WO3 nanoplates for electrocatalytic hydrogen evolution, J. Nanomater. 16 (2015) 23.

[24] A.C. Marques, L. Santos, M.N. Costa, J.M. Dantas, P. Duarte, A. Gonçalves, R. Martins, C.A. Salgueiro, E. Fortunato, Office paper platform for bioelectrochromic detection of electrochemically active bacteria using tungsten trioxide nanoprobes, Sci. Rep. 5 (2015) 9910.

[25] H. Kalhori, M. Ranjbar, H. Salamati, J.M.D. Coey, Flower-like nanostructures of WO3: fabrication and characterization of their in-liquid gasochromic effect, Sensor. Actuator. B Chem. 225 (2016) 535-543.

[26] J. Meng, J. Pei, Z. He, S. Wu, Q. Lin, X. Wei, J. Li, Z. Zhang, Facile synthesis of gC3N4 nanosheets loaded with WO3 nanoparticles with enhanced photocatalytic performance under visible light irradiation, RSC Adv. 7 (2017) 24097-24104.

Chapter 6

Ultrasonic and Spectroscopic Techniques for the Measurement of the Elastic Properties of Nanoscale Materials

Abstract

Materials at the nanoscale often have properties which differ from those they have in the bulk form. These properties significantly depend on the production process, and their measurement is not trivial. The elastic properties characterize the ability of materials to deform in a reversible way; they are of interest by themselves, and as indicators of the type of nanostructure. As for larger scale samples, the measurement of the elastic properties is more straightforward, and generally more precise, when it is performed by a deformation process which involves exclusively reversible strains. Vibrational and ultrasonic processes fulfill this requirement. Several measurement techniques have been developed, based on these processes. Some of them are suitable for an extension towards nanometric scales. Until truly supramolecular scales are reached, the elastic continuum paradigm remains appropriate for the description and the analysis of ultrasonic regimes. Some techniques are based on the oscillations of purpose-built testing structures, mechanically actuated. Other techniques are based on optical excitation and/or detection of ultrasonic waves, and operate either in the time domain or in the frequency domain. A comparative overview is given of these various techniques.

Keywords: elasticity, vibrations, ultrasonics, surface waves, resonators, thin layers, nanorods

1. Introduction

Solid materials exist when the atoms find a configuration in which their potential energy has at least a local minimum, around which a stability region exists. Irrespective of the high number of degrees of freedom, and with rare exceptions, in the neighborhood of any minimum a more or less narrow interval exists, in which the potential energy is well represented by quadratic terms, higher order terms becoming negligible. A quadratic potential energy means an elastic restoring force, meaning that, when considering free motion around a stable equilibrium configuration, vibrational excitations are expected, which have a time periodicity. Periodicity can be represented by a frequency, or an angular frequency ω, which is obviously determined by the properties of the physical system, namely in terms of stiffness and inertia. This general consideration applies from the atomic scale up to the full macroscopic scale, at which continuum models are appropriate.

When looking at the atomic motions, stiffness and inertia are given by the interatomic forces and the atomic masses. A distinction can be proposed among vibrational excitations, based on the phase difference between the motions of different atoms. Excitations exist in which when considering atoms at progressively decreasing distances, down to first neighbors, the phase difference between their displacements does not tend to zero. Examples are the vibrations of molecules, or the optical phonons in a crystalline structure. In this case the average displacement evaluated over a group of neighboring atoms does not have a relevant meaning, since the displacement of single atoms can be significantly different from the average one, and since the average can be null (as in the case of isolated molecules) or close to zero even in presence of atomic motions of significant amplitude. Excitations instead exist in which the phase difference between displacements of atoms located at smaller and smaller distances, down to nearest neighbors spacing, are smaller and smaller. Examples are the acoustic phonons in a crystalline structure. In this case the average displacement evaluated over a group of neighboring atoms becomes fully representative of the displacement of single atoms, and is the natural bridge towards a description of the continuum type, with a displacement vector field $\boldsymbol{u}(\boldsymbol{r},t) = u(x_1,x_2,\ x_3,t)$ which is a continuous function of the position vector $\boldsymbol{r} = (x_1,x_2,\ x_3)$ and of time t. Excitations of this type are called acoustic excitations.

The description by the continuous vector field of displacement, and consequently by the tensor fields of strain and stress, is appropriate at scales which go from the supramolecular one, of the order of the nanometer or slightly more, at which the above-mentioned average begins to be meaningful, up to the fully macroscopic one. In the elastic continuum model [1–3] the potential energy is a quadratic function of strains, the coefficients of the expansion being the elements of the tensor of the elastic constants $\left[C_{ijmn}\right]$, conveniently represented in compact Voigt notation as $[C_{kl}]$. The stiffness being represented by these tensor elements, or by some functions of them which are specific elasticity moduli, and the inertia being represented by the mass density ρ, the equations of motion of the continuous medium are the elastodynamic equations. In the limit of an infinite homogeneous medium, which is symmetric to any translation in position and in time, the fundamental excitations are conveniently taken as waves, which, besides being periodic in time, are periodic in space, as described by a wavelength λ or a wave vector $\boldsymbol{k}$, with $|\boldsymbol{k}| = 2\pi/\lambda$. These fundamental solutions are characterized by dispersion relations $\omega(\boldsymbol{k})$, which are determined by the stiffness and the inertia properties [1–3].

It has been recognized, since long ago, that the measurement of the vibrational excitations gives access to these properties [4]. In particular, if the inertial properties (atomic masses or mass densities) are known, the stiffness properties can be measured. In both the atomic and the continuum case, they represent the curvature of the potential energy in the neighborhood of its minimum, and therefore they contain information, respectively, about the interatomic bonding and about the stiffness of solids, in its usual meaning. A whole wealth of experimental techniques has therefore been developed, which exploit vibrational excitations to measure material properties [4]. A general advantage of all the measurement techniques based on vibrational excitations is that they can exploit displacements of small amplitude, confined in the neighborhood of the equilibrium position, in which the representation of the potential energy by only the quadratic terms is an excellent approximation. In other words, higher order terms of the potential do not interfere, and, in the case of continua, non-elastic deformation mechanisms, other than the simple, reversible, stretching of interatomic bonds, are not activated.

Vibrational excitations of non-acoustic type, which cannot be described by a continuous field of displacements, are not treated by mechanics, but rather by solid state physics, which measures them by techniques like Raman spectroscopy, to obtain various information at atomic level, including that concerning the interatomic forces. They are not considered here.

Vibrational excitations of the acoustic type, instead, have the same properties from the supramolecular, or nanometric, scale, up to the kilometric scale and above. Accordingly, techniques based on acoustic excitations are exploited to measure the stiffness properties of objects of various sizes, up to dams, and are exploited for geological investigations. These methods measure the dynamic, or adiabatic, elastic moduli; these moduli do not coincide with the isothermal moduli which are measured in monotonic tests (if strain rate is not too high), but in elastic solids the difference between adiabatic and isothermal moduli seldom exceeds 1% [2].

At the other extreme of the size scale, some techniques can be pushed to measure microscopic objects, down to carbon nanotubes. Nanomechanics precisely addresses the behavior of microscopic objects; in this chapter we consider measurement techniques based on vibrational excitations of the acoustic type, which are pushed to measure small objects, towards the size at which the same concept of 'acoustic excitations' begins to lose significance, as well as that of the strain field. Techniques based on vibrational excitations, and in particular optical techniques, which avoid mechanical contact, are particularly prone to be applied to small objects, for which the contact with actuators or sensors becomes critical. In many cases, the mechanical properties of materials at this scale are of interest for the design and production of microsystems which operate dynamically; in these cases, the dynamic, adiabatic, moduli, are precisely those of interest.

The next section summarizes some basic concepts about free acoustic excitations in finite objects. The following two sections give an overview of several measurement techniques, which are grouped in two categories. First, those that exploit the oscillations of purpose-built testing structures, which are mechanically actuated, often by piezoelectric means. Secondly, the techniques which measure the properties of ultrasonic waves, that are typically excited and/or detected by optical means. These techniques can be further subdivided among those which operate in the time domain and those which operate in the frequency domain.

2. Acoustic excitations in confined media

Displacements and strains in the elastic continuum model in the absence of body forces, obey the elastodynamic equations; for homogeneous media they are [1–3]

$$\rho \frac{\partial^2 u_i}{\partial t^2} = \sum_{j,m,n} C_{ijmn} \frac{\partial^2 u_m}{\partial x_j \partial x_n}, i = 1, 2, 3.. \tag{1}$$

The invariance to any translation in time is the root of the harmonic time dependence of the fundamental solutions, characterized by the circular frequency ω, which allows to transform the equations into the Helmholtz equations. The invariance to any translation in position, in practically infinite media, is the root of the harmonic space dependence of the fundamental solutions, which are traveling monochromatic harmonic waves, characterized by the wave vector $\boldsymbol{k}$. The dispersion relations $\omega(\boldsymbol{k})$ are determined by the properties of the medium. The displacement vector $\boldsymbol{u}$ having three independent components, the dispersion relation has three branches, which can be classified according to the relative orientations of the vectors $\boldsymbol{u}$ and $\boldsymbol{k}$, i.e. according to their polarization: in an isotropic medium, one longitudinal and two transversal modes [1–3].

The elastic continuum model has no intrinsic length scales; as mentioned above it loses significance at the nanometric scale, while in an infinite medium, which also from the geometrical point of view has no intrinsic length scale, it does not have an upper limit of size. The wavelengths span in a continuous way this whole infinite interval, and the corresponding angular frequencies go, in continuity, from null frequencies for $\lambda \to \infty$, i.e. $\boldsymbol{k} \to \mathbf{0}$, up a not sharply defined upper limit, corresponding to the shortest meaningful wavelengths. The absence of intrinsic length scales implies that all these wavelengths behave in exactly the same way, and the dispersion relations are simply linear: $\omega = v|\boldsymbol{k}|$, the velocities being independent from $|\boldsymbol{k}|$, i.e. these modes are non dispersive. In the general anisotropic medium the velocities depend on the direction of $\boldsymbol{k}$, while in the isotropic case they do not, and the velocities v_l and v_t of the longitudinal and transversal modes are respectively [1–3]

$$v_t = \sqrt{C_{44}/\rho} \text{ and } v_l = \sqrt{C_{11}/\rho}, \tag{2}$$

thermodynamic stability requiring that $v_t < v_l$.

Rupture of the unconditional translational symmetry by some kind of boundary condition, which introduces some kind of confinement, induces the appearance of further acoustic modes, namely standing waves. The consequences of confinement can be appreciated also without abandoning the relative simplicity of the isotropic model. They are already present, in a paradigmatic way, in the case which is probably the simplest rupture of the infinite translational symmetry: the plane external surface of a semi-infinite medium. The invariance to any translation in time is not altered, and correspondingly the acoustic modes remain periodic in time, associated to a circular frequency ω. In the same way, the invariance of the medium to any translation in the plane of the surface is not modified, and correspondingly the acoustic modes remain periodic, and traveling, in this plane. This periodicity is conveniently represented by a wave vector $\boldsymbol{k}_\parallel$ parallel to the surface, which identifies a direction and a repetition period. Considering media which possess in-plane isotropy, the direction of $\boldsymbol{k}_\parallel$ becomes irrelevant and, in order to simplify the notation, we introduce here the symbol $\lambda_\parallel \equiv 2\pi/|\boldsymbol{k}_\parallel|$, which seems inappropriate because wavelength is not a vectorial quantity, but must be understood as a compact form of: "the period along the direction of $\boldsymbol{k}_\parallel$ of an acoustic excitation whose wave vector component is $\boldsymbol{k}_\parallel$".

In the direction perpendicular to the planar external surface (the direction of depth), two types of space dependence are instead found. A set of modes takes advantage of the (semi) infiniteness of the medium, and maintains a space periodicity, conveniently represented by a wave vector $\boldsymbol{k}_\perp$ perpendicular to the surface. These modes are thus characterized by a full three dimensional wave vector $\boldsymbol{k} = \boldsymbol{k}_\parallel + \boldsymbol{k}_\perp$, and are completely analogous to those of an infinite medium: they are the bulk waves, which are reflected at the surface. There is no upper limit to their wavelengths, and no lower limit to their frequencies. The only novelties are introduced upon reflection: firstly, in the direction perpendicular to the surface the interference between the incident and the reflected waves generates a standing wave pattern. Secondly, the mere superposition of an incident longitudinal wave and its reflected counterpart would not satisfy the stress free boundary conditions. Therefore, upon reflection, an incident longitudinal wave is partially reflected into a longitudinal wave, having the same $\boldsymbol{k}_\parallel$ and a reversed component $\boldsymbol{k}_\perp$, and partially converted into a transversal wave, having the same frequency and the same $\boldsymbol{k}_\parallel$, but a different value of $\boldsymbol{k}_\perp$, because it has a different velocity [1, 3]. The dual conversion occurs for an incident transversal wave.

However, in the presence of this boundary, the elastodynamic equations admit another set of solutions, which are not periodic in the direction perpendicular to the external surface. They are the surface acoustic waves (SAWs), of which the

Rayleigh wave, the only one existing at the free surface of a semi-infinite homogeneous medium, is the prototype. The Rayleigh wave has a displacement field which decays exponentially with depth, with a decay length which is uniquely determined by the elastodynamic equations, and turns out to be very close to $\lambda_{\parallel}$; it is thus confined in the neighborhood of the external surface [1, 3, 5]. The Rayleigh wave also has another character typical of the modes induced by confinement: they do not have a specific polarization. The displacement vector $\boldsymbol{u}$ of the Rayleigh wave has a direction which changes with depth; it always is in the plane identified by $\boldsymbol{k}_{\parallel}$ and by the normal to the surface, which is called the sagittal plane, but the two components vary with depth in different ways. The Rayleigh wave has its own velocity v_R, lower than that of any bulk wave: $v_R < v_t < v_l$ [1, 3, 5]. Acoustic modes having similar properties can also appear at the surface separating two media in perfect adhesion.

In the semi-infinite case the medium still has no intrinsic length scale. Both periods, $\lambda_{\parallel}$ and $2\pi/|\boldsymbol{k}_{\perp}|$, can go in continuity from infinity to the smallest meaningful values, and correspondingly the frequencies go from zero to a very high upper limit. Accordingly, the bulk waves and the Rayleigh wave are non-dispersive, also the velocity v_R being independent from $|\boldsymbol{k}_{\parallel}|$; the dispersion relation $\omega = v_R|\boldsymbol{k}_{\parallel}|$ is linear. However, for any value of $\boldsymbol{k}_{\parallel}$, the value of $\lambda_{\parallel}$, which determines the decay length, somehow sets a length scale. It can approximately be said that the value of v_R is mainly determined by the properties of the medium up to a depth $\lambda_{\parallel}/2$; in other words, the Rayleigh wave is a probe which senses the properties of the medium up to that depth [6]. The component $\boldsymbol{k}_{\perp}$ exploring the whole interval from zero to the maximum meaningful value, we have $|\boldsymbol{k}| = |\boldsymbol{k}_{\parallel} + \boldsymbol{k}_{\perp}| \geq |\boldsymbol{k}_{\parallel}|$, and λ cannot be below $\lambda_{\parallel}$. For that value of $\boldsymbol{k}_{\parallel}$ we have bulk modes, whose frequencies go with continuity from $v_t|\boldsymbol{k}_{\parallel}|$ till very high values, and, below this lower limit, the Rayleigh wave at an isolated frequency $v_R|\boldsymbol{k}_{\parallel}|$.

Instead, in media that are finite in at least one dimension, with a size D, the size of the object which supports the excitations sets a reference length scale. The relevance of confinement is determined by the ratio of the wavelength λ to the size D. On the high side, acoustic excitations having a ratio λ/D significantly larger than one are not supported: the size D sets an upper limit for wavelength. On the lower side, the excitations having a ratio $\lambda/D \ll 1$, down to the lower limit at which the continuum model loses significance, are affected in a negligible way by the finiteness of the medium. For these excitations, the medium is still almost invariant for translations of several wavelengths, and the modes are indistinguishable from traveling periodic waves. However, strictly speaking, they become standing waves, and the difference becomes evident for the excitations whose ratio λ/D is not far from unity. For these excitations, the medium is definitely not invariant for translations of few wavelengths: the modes cannot have a well-developed periodicity, and are strongly affected by confinement.

Objects having a high aspect ratio, like thin layers or beams, or nanorods, are often of interest: in one or two dimensions their size is D, in the remaining direction(s) it is D', with $D \ll D'$. The consequences of confinement within D are evident for excitations having $\lambda \sim D$, well before the finiteness of D' becomes perceptible. These excitations can be analyzed as if in the other direction(s) the medium was still infinite: the excitations are still periodic in this (these) direction(s), characterized by a wave vector $\boldsymbol{k}_{\parallel}$. The effects of the finiteness of D' become evident only when $\lambda_{\parallel} \sim D'$.

Again, the Rayleigh waves paradigmatically indicate the consequences of a finite size. We can consider a slab, whose thickness is the characteristic size D, and which can be considered infinite in the other directions. Bulk waves with $\lambda/D \ll 1$ are affected in a negligible way by the finiteness of the medium, as well as the SAWs having $\lambda_{\parallel} \ll D$. At such wavelengths, the Rayleigh waves are confined in the neighborhood of the two surfaces, and do not interact with each other, while the bulk waves have a component $\boldsymbol{k}_\perp$ which is now discretized (the period $2\pi/|\boldsymbol{k}_\perp|$ must be an integer sub-multiple of D) but with a narrow spacing. Instead, when λ increases and approaches D, on one side this discretization becomes relevant, and, on the other side, the tails of the displacement fields of the Rayleigh waves at the two surfaces superpose. The two SAWs then merge into modes which are typical of the slab, with displacement fields which extend throughout the thickness, their depth dependence being not periodic. Among them, the bending modes of the slab [7, 8], reminiscent of the bending modes of a membrane, which are different from the transversal bulk modes, which are shear modes, not bending modes. The dispersion relation $\omega = \omega(|\boldsymbol{k}_\parallel|)$ has discrete branches which correspond to these modes. They are nonlinear, i.e. dispersive, meaning that their velocities depend on the period $\lambda_\parallel$, more precisely on the ratio $\lambda_\parallel/D$, i.e. on the product $D|\boldsymbol{k}_\parallel|$; they asymptotically tend to the linear dispersion relations of the infinite medium when the product $D|\boldsymbol{k}_\parallel|$ becomes large.

In the case of a film supported by a substrate, which can be generalized to stratified media, the layer thickness sets the characteristic size D. Bulk waves in any layer, with $\lambda/D \ll 1$, and the Rayleigh wave at the surface of the outermost layer with $\lambda_\parallel \ll D$, are indistinguishable from those in a semi-infinite medium of the same material. Instead, waves with λ, or $\lambda_\parallel$, comparable to, or larger than D (if the substrate is semi-infinite λ can go to infinity) have a displacement field which extends over various layers, and have properties which depend on the properties of the various layers. In particular, the Rayleigh wave becomes a generalized Rayleigh wave, whose depth dependence is affected by the transition from the film to the substrate. When $\lambda_\parallel \gg D$ the decay length is also much larger than the film thickness, and the displacement field of the generalized Rayleigh wave is mostly in the substrate. In this limiting case the generalized Rayleigh wave approaches the Rayleigh wave of the bare substrate, only slightly modified by the presence of the supported layer. Depending on the properties of the layers, namely on their acoustic velocities, other SAWs can be supported. For instance, an acoustically slow layer can act as a waveguide, confining some other modes, like the Sezawa modes. These modes are reminiscent of the modes of a slab, but instead of having stress free boundary conditions they have, on the substrate side, continuity boundary conditions, and tails of the displacement field which extend into the adjacent layers. Obviously also these modes are dispersive, their velocities depending on the product $D|\boldsymbol{k}_\parallel|$, or, in the case of several layers, on the products with the various thicknesses. The dispersion relations for the various branches can be numerically computed, as functions of these products and of the elastic constants and mass densities of all the layers [9, 10].

A wide slab, of thickness D, can be cut into a stripe of width comparable to D: confinement thus occurs in two directions. More generally, it is the case of a slender cylinder, of circular or non-circular cross section, of lateral size D, which can still be treated as infinite in the third direction. The parallel wave vector $\boldsymbol{k}_\parallel$ remains fully meaningful, although, having exclusively the axial direction, its vectorial character becomes redundant. New modes appear, namely the torsional ones, beside the bending or flexural ones, and the dilatational ones. In the case of circular cylinders they obey the Pochammer-Cree Equations [11, 12], in which, as it often happens in cylindrically symmetric cases, Bessel functions play a crucial role. More general cases can be analyzed by the so-called xyz algorithm [13, 14], which has been applied to rectangular [15], circular [16] and hexagonal [17] cases, and also to more complex cases like superlattices [18].

The dispersion relation $\omega = \omega(|\boldsymbol{k}_\parallel|)$ has a significant number of branches, which asymptotically tend to the linear dispersion relations of the infinite medium when the product $D|\boldsymbol{k}_{\parallel}|$ becomes large. For smaller values of $D|\boldsymbol{k}_{\parallel}|$, instead, these branches remain separate, and nonlinear. In particular, several branches have a non-null frequency for $|\boldsymbol{k}_{\parallel}| = 0$; they correspond to modes, like e.g. the radial breathing modes, which can have a non-traveling character [12]. Furthermore, for small values of $D|\boldsymbol{k}_{\parallel}|$, the dispersion relations of most of these modes have a slope much smaller than the velocities v_t or v_l, meaning a low group velocity, and in some cases even a negative slope [19].

Finally, for objects whose aspect ratio is not far from unity, and therefore in which confinement is along all the three directions to a size of the same order D, the general considerations apply. Namely, all the excitations have the character of discrete standing waves. For ratios $\lambda/D \ll 1$ the discretization has a narrow spacing, the modes are almost indistinguishable from those of an infinite medium, and can still be characterized by a wavevector $\boldsymbol{k}$, or $\boldsymbol{k}_{\parallel}$. Instead, for ratios λ/D not far from unity the modes, in general, do not have a regular periodicity, and any wavevector loses its meaning. The modes have clearly separate frequencies, are strongly affected by confinement, and strongly depend on the shape of the object. Note that for objects which become nanometric the regime in which $\lambda/D \ll 1$ does not exist, because it would mean wavelengths at which the continuum model breaks down.

Two main geometries are of interest, both for fundamental studies and for technological applications. Firstly, the planar geometry of thin films, in which confinement is in only one direction, the critical size D is the thickness, the wider size D' being the lateral extension of the layer. The typical examples are the supported films and the resonators, either in the form of a clamped membrane or of a cantilever, or a bridge. Secondly, the linear geometry of beams and of nanorods, in which confinement is in two directions, the critical size D is the diameter, the wider size D' being the length. The two characteristic lengths, D and D', identify two length ranges; acoustic excitations can be probed at the two length scales, which obviously correspond to two frequency ranges.

Ultrasonic waves of wavelength comparable to D are not affected by the finiteness of D', or by the precise shape of the object supporting them: they see the other dimensions as infinite. The properties of the waves depend on the material properties, and the value of D determines the existence of the discrete modes, which can be observed. In some cases the properties of the waves are also affected by the value of D, as it happens for supported films when the ratio λ/D is close to unity: in this case the displacement field of the wave appreciably penetrates into the substrate, and is affected by its properties.

Instead, for acoustic excitations at the scale of D', the effective stiffness for the vibrational modes depends on the material properties, but also, crucially, on the value of D. The properties of the waves thus depend on the material properties and on the value of D, while the value of D' determines the existence of the discrete modes, which can be observed. In fact the fundamental, and higher order, oscillations of a cantilever can be seen as standing flexural waves, of wavelength comparable to of D', and of effective stiffness dependent on D.

The two ranges of wavelength identify two classes of measurements. A first class exploits purpose-built testing structures, which determine the value of D'; they are probed at effective wavelengths of this order. Typically, their oscillations are mechanically actuated by piezoelectric means. The next section is devoted to them. A second class exploits ultrasonic waves at wavelengths comparable to D. Their excitation is achieved by short laser pulses, or simply by thermal motion. Their detection typically requires optical means, and can be performed either in the time domain or in the frequency domain.

3. Measurement techniques based on testing structures

As repeatedly underlined, the elastic continuum model is meaningful down to almost the molecular scale. Techniques based on acoustic excitations are therefore, in principle, applicable to objects down to the nanometric scale. Their effective application depends on the availability of appropriate transducers to excite and to detect the relevant acoustic excitations. In the case of macroscopic objects, the typical techniques are based on piezoelectric transducers, which can be exploited for both excitation and detection. Specific devices are available, and specific instruments, like acoustic microscopes. An alternative, for what concerns detection, is offered by laser Doppler vibrometry. It has the advantages of the optical techniques: light is a massless probe, contactless, which does not load the measured object, is free from own resonances, has a bandwidth that is essentially determined by electronics (the light sensor and the amplification). Furthermore, it can measure small objects, and can measure surfaces which are difficultly accessible, or on which the application of a detector is not possible, e.g. because of their temperature.

In the case of small objects, down to micrometric or sub-micrometric scale, the exploitation of a separate measurement device in mechanical contact becomes impossible. Mechanical actuation by piezoelectric means is still possible either in the case of resonators, by an external actuator which shakes the whole assembly containing the resonant structure, or by inclusion, by nanofabrication techniques, of a piezoelectric (nano)layer as an integral part of the structure being tested. MEMS/NEMS of this type can also be actuated by electrostatic means. In some cases, the test system can have a structure analogous to that of a complete device, and comprehend, beside the possibility of excitation by piezoelectric or electrostatic means, the ability of measurement, e.g. by capacitive detection or by an interdigitated transducer (IDT). Except for these cases, optical detection is mandatory. Laser Doppler vibrometry is the measurement technique of choice, if the measured system has a flat surface of sufficient size. Otherwise, interferometric techniques have been exploited.

A specific case is that of resonators: a significant effort is under way to produce high quality resonators, mainly because of their great potential for applications in sensing, signal processing, and quantum physics. The properties of a resonator can be measured in a static way, by measuring the deflection of a cantilever, or a membrane, or by measuring its resonant frequency. When a resonator is reduced to a small size, typically in the shape of a cantilever, a bridge, or a clamped membrane, its surface to volume ratio increases. Therefore phenomena, which otherwise are minor or negligible, and are not accounted for by q. (1), become non negligible; namely anelastic effects, possibly connected to internal friction phenomena, and surface phenomena, like surface tension or environmental effects [20, 21]. In particular, interaction with the environment, typically by adsorption of molecules, including water, made available by relative humidity, is the physico-chemical basis for the development of sensors. The development of high performance sensors based on nanoresonators [22, 23] is not considered here. We merely note that research in this direction has led to doubly clamped resonators of thickness down to 22 nm and aspect ratio up to 5000; the measurement of their deflection requires an interferometric technique, and their motion due to Brownian thermomechanical techniques becomes detectable [24].

However, resonant structures can built specifically to the purpose of a precise measurement of the properties of the materials which constitute them. In order to measure the properties of tetrahedral amorphous carbon (also known as diamond-like carbon), Czaplewski et al. built, by standard techniques for the production of

MEMS, several resonators, with critical dimension down to 75 nm [25]. Excitation was in some cases electrostatic, and in other cases by shaking by an external piezoelectric actuator. Detection was, depending on the in-plane or out-of-plane deflection, by a laser deflection technique similar to that used by AFMs, or by an interferometric technique. Their results allowed them to analyze dissipation and to discuss various possible mechanisms. Exploiting flexural and torsional oscillators, and the same interferometric measurement technique, they also determined the elastic moduli as function of temperature [26]. In order to measure the properties of an assembly of only carbon nanotubes (called 'forest of nanotubes'), self-sustained only by their entwining and internal interactions, Hassan et al. built by this material cantilevers, of 1 mm length, which were electrostatically excited, and whose motion was detected by laser Doppler vibrometry [27].

The advantages of miniaturization led to explore also the sizes at which the elastic continuum approach is no longer adequate, and a Molecular Dynamics approach is more appropriate [28]. Interestingly, for membranes which become nanometric but have significantly larger lateral extension, the continuum approach remains useful, both for the theoretical analysis and the measurement technique. Membranes of nanocrystalline diamond and of piezoelectric aluminum nitride, and bilayer membranes, with thicknesses down to 220 nm and diameters up to 1 mm, have been investigated by Knoebber et al. [29] by both a static and a vibrational technique. The static technique, of more macroscopic character, was the bulge test, in which the deflection of the circular clamped membrane under gas pressure is measured; the implementation was optical, the deflection being optically measured by white light interferometry. The vibrational technique involved excitation by an external piezoelectric stack, and detection by laser Doppler vibrometry. The analysis shows that the results from the dynamical technique are less sensitive to the geometrical inaccuracies of the tested membrane. Similarly, the properties of bilayers obtained by growing nanocrystalline diamond on aluminum nitride thin films, of about 200 + 200 nm thickness, were measured producing microresonators, either cantilevers or bridges, of length up to 50 μm, piezoelectrically actuated and measured by laser Doppler vibrometry [30]. Similar nanocrystalline diamond/aluminum nitride membranes, of thicknesses of the order of hundreds of nanometers, were still characterized by the bulge test on circular clamped membranes, of radii up to 1 mm [31].

Also in the analysis of a typical 2-D material, MoS_2, resonators have been built in the form of clamped membranes, of radii of 2 and 3 μm, ranging from a single layer, i.e. a truly atomic thickness, up to over 90 layers. The membranes were obtained over pre-patterned circular holes in a Si substrate, and acted as 'drums'. Their motion was measured exploiting the vibrating drum membrane and the bottom of circular hole as the two mirrors of an interferometer. The continuum mechanics approach turned out to be useful in the interpretation of the experimental results, which showed the transition from the membrane regime, in which the restoring force in the oscillation is supplied by the membrane tension, to the plate regime, in which the restoring force is supplied by the bending stiffness of the plate [32].

A whole class of devices exploits the intergiditated transducers (IDTs) to launch and resonantly detect surface acoustic waves. A piezoelectric layer is a crucial component of these devices. When the lithographic techniques to produce IDTs is available, devices have been produced specifically for the aim of measuring the properties of the material which constitute them. Measurements have been performed on various forms of artificial diamond (nanocrystalline diamond, nitrogenated diamond-like carbon), of particular interest for devices exploiting surface waves and IDTs because of their high acoustic velocity [33–35].

4. Measurement techniques based on ultrasonic waves

Various measurement techniques based on vibrations and acoustic waves rely on impulsive mechanical excitation of vibrational modes or of waves. Mechanical excitation by an impact has been analyzed in detail [36, 37], and standards have been issued concerning it [38]. When extending these techniques towards nanomechanics, and in particular to thin films, wave excitation by laser pulses is their natural evolution. Absorption of a laser pulse, of duration τ, induces, by thermal expansion, a sudden expansion: a strain pulse, which propagates away at the speed v_l of longitudinal sound. The mechanism of phonon generation has been analyzed in detail [39]. The geometry of this pulse depends on the absorption length ζ of the optical pulse, on the distance $v_l\tau$ traveled by the strain pulse during the laser pulse, and on the lateral size of the region on which the laser beam is focused. Metals have the shortest absorption lengths, of some nanometers, and have longitudinal sound velocities of few km/s, i.e. few μm/ns. With optical pulses of the order of the ns, heating and thermal expansion occur until the pulse has traveled a distance of few μm.

In the case of a thin supported film, of thickness D of some micrometers, or even less, a ns laser pulse launches a strain pulse which is originated in the whole thickness of the film. Such a strain pulse travels parallel to the surface, and, if the film has a sufficient lateral extension, can be detected at a distance of millimeters, measuring the transit time. This is the regime of the so-called laser ultrasonics, in which the laser is focused by a cylindrical lens. A line source is therefore present in the film, which launches two SAWs traveling in opposite directions, on a surface whose lateral size D' is large. The ratio $\lambda_{\|}/D$ can be smaller than one, implying that the displacement field is essentially confined within the film, or larger, meaning that the displacement fields of the SAWs penetrate significantly into the substrate. In the latter case a typical objective of the measurement is the detection of the (small) modification of the Rayleigh wave of the bare substrate.

With optical pulses of less than the picosecond, instead, heating and thermal expansion occur only until the strain pulse has traveled a distance of the order of the nm, i.e. less than the absorption length. The pulse displacement during excitation is thus negligible, and heating, and thermal expansion, occur only in a thin surface skin, of depth ζ, i.e.few nm. This is the regime of the so-called picosecond ultrasonics, in which the laser is focused by a spherical lens to a spot whose width is of several micrometers, i.e. orders of magnitude wider that the depth within which thermal expansion occurs. In the case of a film of thickness D, a plane wave is therefore launched: a strain pulse, of lateral extension of several μm, which extends over a depth of ζ, has leading wavelength of this same order, and travels in the direction perpendicular to the surface. Except for truly nanometric films, in this case $\lambda/D \ll 1$, and the finiteness of D becomes relevant thanks to the reflection of the pulse when it reaches the opposite surface. If the film is supported the reflection is only partial, part of the pulse being transmitted; the echo however returns to the surface, where it can be detected, and where it is again reflected back. The transit time, of the order of the ns, or less, is measured by a pump-and-probe technique, in which a short, weaker, 'probe' pulse follows the 'pump' pulse with a variable delay, and detects the time evolution of the transient induced by the pump pulse, scanning the delay until the exhaustion of the transient itself.

The technique has also been exploited for nanostructures which do not have a planar surface extending over the several micrometers of the focused spot; nanorods are an example. In this case the geometry of the strain pulse is not the simple planar one outlined above, and is rather dictated by the specific geometry of the illuminated object(s). The detected signal contains however relevant information, whose interpretation typically requires the modeling of the specific dynamic structure of the investigated objects.

In both the above techniques a transient is induced by a pulse which is short, i.e. strongly localized in time, and also strongly localized in space, resulting in a broadband and localized pulse, whose transit is easily detected. In a different technique a similar short pulse is exploited, which produces a transient, but illuminates a wider segment of the surface with a periodic pattern, and has therefore a narrow band. Periodicity is obtained splitting the pump laser pulse in two beams, and recombining them at the surface of the sample, forming an interference pattern. The same impulsive thermal expansion thus occurs with a periodic space modulation. The periodicity, which here is called $\lambda_{\|}$ and which in the literature dealing with this technique is more often called Λ, is selected by the interference geometry, as well as its direction: the wavevector $\boldsymbol{k}_{\|}$ is selected by the illumination geometry. The technique is applicable to samples having a planar surface whose lateral size D' is much larger than the periodicity $\lambda_{\|}$.

The time evolution of the reflectivity (or, possibly, of the transmittance) is measured in one point, either by a continuous measurement by a fast detector, or, more frequently, by the pump-and-probe technique. The measured signal typically has a slowly declining component, corresponding to the total energy deposited by the pulse, which eventually diffuses towards the adjacent parts of the sample, and a periodic component. The periodic component is due to the SAWs which are launched in the directions of $\boldsymbol{k}_{\|}$ and $-\boldsymbol{k}_{\|}$, and which form a standing wave, eventually damped, both because the two SAWs travel away, and because of intrinsic damping mechanisms, like the thermoelastic one. By the spectral analysis of the observed signal one point of the dispersion relation $\omega = \omega(\boldsymbol{k}_{\|})$ is obtained, hence the name of 'transient grating spectroscopy'. The full dispersion relation is obtained scanning the wavevector $\boldsymbol{k}_{\|}$, by adjusting the geometry of the interfering beams; at least in principle, a wide interval of $|\boldsymbol{k}_{\|}|$ is accessible. In the case of a film of thickness D, the interval from $\lambda_{\|}/D \ll 1$ to $\lambda_{\|}/D \gg 1$ can sometimes be scanned. It must however be noted that this is a case in which the characteristic length D is dictated, more than by the geometry of the sample, by the geometry of the excitation, i.e. by the periodicity $\lambda_{\|}$ determined by the experimental set-up. In fact, as noted above, SAWs of wavevector $\boldsymbol{k}_{\|}$ are sensitive to the properties of the medium only up to depth of approximately $\lambda_{\|}/2$.

In all the above techniques a 'pump' laser pulse induces a transient, whose time evolution is measured, with a time resolution down to the picosecond scale. The transient grating technique deserves the 'spectroscopy' name because it relies on the successive spectral analysis of the measured time signal. Instead, the vibrational spectroscopies investigate the steady state excitation of vibrational modes, due to thermal motion. They exploit continuous, not pulsed, lasers, and detect the component of the scattered light which has undergone a frequency shift, because it was inelastically scattered by the excitations present in the sample. Obviously, inelastic scattering involves some energy exchange between the optical and the acoustic fields, but, due to the smallness of the scattering cross section, this is minor, in comparison to the thermal energy. Only in special cases, namely highly confined nanostructures, the exchange can become significant. The most widespread vibrational spectroscopies, namely Raman spectroscopy and infrared spectroscopy, are not considered here because they measure vibrational excitations of non acoustic type.

Instead, Brillouin scattering is precisely the inelastic light scattering by the vibrational excitations of the acoustic type, and Brillouin spectroscopy measures it.

In a manner analogous to that of Raman spectroscopy, the sample is illuminated by a laser beam, and the scattered light is collected; the spectral analysis singles out the minor fraction which has undergone inelastic scattering by the excitations in the sample. The scattering geometry selects an exchanged wavevector $\boldsymbol{k}$, or $\boldsymbol{k}_{\parallel}$ in the case of SAWs. Therefore, also in this technique, in the case of SAWs the experimental technique selects a periodicity $\lambda_{\parallel}$ which essentially dictates the length scale D, and the depth over which the properties of the medium are interrogated. A Brillouin spectrum supplies a point of the dispersion relation $\omega = \omega(\boldsymbol{k})$ for each of the branches which give a measurable peak. With visible light, and typical properties of solid materials, the observed frequencies, i.e. the frequency shifts of light, range from few GHz to tens of GHz, i.e. a wavenumber in the range of the cm^{-1}, to be compared with the typical frequency shifts observed in Raman spectroscopy, in the range of hundreds of cm^{-1}. Accordingly, the diffraction gratings, which are the common spectrum analysers in Raman spectroscopy, do not have a sufficient resolution, and other techniques must be employed. The tandem multipass Fabry-Perot interferometer has become the standard apparatus in Brillouin spectroscopy [40]. The small observed frequencies (1 cm^{-1} means 30 GHz or 1.4°K) make the Stokes and anti-Stokes parts of the spectra symmetric, to a difference from what happens in Raman spectroscopy.

The dispersion relations $\omega = \omega(\boldsymbol{k})$ of the traveling waves, whenever a wavevector $\boldsymbol{k}$ can meaningfully be identified, and the frequencies of the standing waves, can be theoretically predicted, from the simplest ones of Eqs.(2) to the more complex ones which can only be numerically computed. They are all functions of the properties of the medium (or the media involved). Therefore, whenever and however they are measured, they provide access to the properties, which can be found fitting the computed dispersion relations, or the frequencies, to the measured ones. Obviously the amount of obtainable information depends on the amount, and the quality, of the experimental evidence, and on the simpler or more sophisticated way of treating it. The obtainable information ranges from a semi-quantitative comparison for a single parameter, to a complete elastic characterization of a layer.

4.1 Laser ultrasonics

The laser ultrasonics technique exploits SAWs, mainly the Rayleigh wave, possibly modified by the presence of a supported film, to measure the properties of samples having a planar surface of sufficient lateral extension. The focus often is the measurement of the properties of the supported film. Waves are launched by a laser pulse, typically of nanosecond duration; visible or near UV light is typically used, often from a N_2 laser at 337 nm. As mentioned above, the laser is focused by a cylindrical lens on the surface of the sample, resulting in a sudden expansion of a line-shaped region, which launches a broadband surface wave, which propagates perpendicularly to the focusing line, with a limited divergence. The surface wave is detected by the displacement it induces perpendicularly to the surface, after a propagation path, typically of millimeters.

Detection can be done by optical interferometry [41, 42], or, in a simpler way, by piezoelectric sensing: either a piezoelectric polymer foil pressed on the sample by a blade [43–47], or a piezoceramic stripe [48]. A ready-to-use commercial apparatus is also available.

Yang et al. implemented both the optical and the piezoelectric detection. They performed a systematic comparison exploiting a 320 nm SiO_2 thermal oxide layer over the pristine Si substrate. The optical detection obtains lower intensity signal but also a lower background noise, and wider bandwidth: in their implementation the bandwidth of the piezoelectric detection is limited to about 120 MHz, mainly by the piezoelectric foil exploited for transduction, while the optical detection shows substantial signal components up to almost 300 MHz [49]. The same authors then integrated the two techniques in a single apparatus.

The recorded displacement can be frequency analyzed, yielding the dispersion relation $v_R(\omega)$ for a frequency interval that can extend over a full frequency decade (e.g. 20 to 200 MHz). As mentioned above, fitting the computed dispersion relation to the measured one allows to derive the film properties. In the case of supported films the width of the measured frequency interval can allow to appreciate the non linearity of the velocity dependence of the frequency. If this is the case, the fit of the computed dispersion relation to the measured one allows to derive both the Young modulus and the film thickness [44, 45]. If the measurement extends over a more limited frequency interval, or if the material properties and the thickness are such that the non linearity is mainly in a frequency interval external to the measured one, an independent measurement of thickness is needed. Since the waves are detected after a propagation of the order of millimeters, the obtained properties are representative of an average of the properties over this distance.

The technique has been extensively adopted to characterize diamond-like carbon films. It has been pushed to the measurement of films having thickness down to 5 nm, deposited on a Si substrate [41]. The stretch of the observed propagation path to 20 mm allowed to measure variation of the Rayleigh wave velocity below 0.25 m/s, over a velocity of about 5000 m/s for the bare Si substrate. The small variation due to the nanometric film is thus detected. Ultra nanocrystalline diamond films of micrometric thickness were more easily characterized [47]. Also the commercially available apparatus proved to be able to measure the properties of alumina films produced by the atomic layer deposition (ALD) technique, of thickness down to about ten nanometers [50].

4.2 Picosecond ultrasonics

Since femtosecond laser pulses became available, the so-called 'picosecond ultrasonics technique' exploits them; nevertheless, it is still called 'picosecond ultrasonics' from the picosecond, or sub-picoseconds, pulses which were available at the time in which it was first demonstrated, by the seminal work of Thomsen et al. [51, 52] and by Wright [53, 54]. It is a technique which belongs to the wide family of the optical pump-and-probe scheme, whose performances, namely resolution, crucially depend on the short duration of pulses.. As previously outlined, two length scales contribute to determine the shape of the ultrasonic field generated by the pump pulse: the absorption length ζ of the optical pulse, i.e. the depth within which the pulse energy is deposited, and the displacement $v_l\tau$ of the strain field in the time interval within which the energy is deposited.

Lasers in the near infrared are typically adopted, to avoid possible spurious effects, which were attributed to electronic interband transitions [55], which might be induced at shorter wavelengths. Typical metals properties correspond to absorption lengths of nanometers; aluminum is among the metals which have the shortest absorption lengths, and well absorbs at 810 nm wavelength [56]. With the typical acoustic velocities and the typical thermal diffusivities of metals, with pulse lengths below the picosecond the fraction of the pulse energy which is not reflected is deposited before both the strain pulse and the temperature rise leave the absorption length. Therefore, the (over)simplified picture can be given, according to which at the end of the pump pulse no significant motion has occurred yet, and a significant temperature rise and dilatational strain are present within an outer skin whose thickness is of the order of the absorption length ζ. The laser being focused on a spot whose width is of several, up to few tens, micrometers, if the outer surface of the sample is planar the strain and the temperature fields are essentially uniform across the spot, and a plane wave is launched in direction perpendicular to the surface. The pulse is localized within a depth interval of the order of 2ζ; it crosses the layer, is reflected (only partially if the film is not free standing), and returns to the surface, where it is detected. Further round trips can also be detectable.

Detection is performed by the probe pulse, much weaker than the pump pulse, which follows it with a variable delay, controlled by a delay line. As it is typical for pump-and-probe techniques, several details of the experimental procedure are tuned to single out the reflection of the probe beam, like different polarizations of the pump and probe beams, or frequency doubling of the probe beam, and to detect small variations of the reflectivity, like differential or interferometric detection [56]. The reflectivity is modified by the strain pulse, via the elasto-optic effect: the strain modulates the optical properties of the film, both the real part and the imaginary part of the refractive index. The measured reflectivity has a slowly varying background, due to the diffusion of the heat deposited by the laser pulse towards the depth of the sample. Superposed to this background, short modulations of the reflectivity denote the arrival of the echoes at the outer surface.

The arrival of the echo(es) at the surface is thus detected, the delay providing a measure of the velocity of the longitudinal acoustic wave, for propagation in the direction normal to the surface. Some details of the detected pulse can, at least in principle, supply further information about the film and the mismatch of properties between the film and the substrate [56]. Among them, the so-called Brillouin oscillations, which are due to the interference between the fraction of the probe pulse reflected at the outer surface, and the fraction which is reflected by the traveling strain pulse. In fact, the strain pulse, which extends over a depth interval of the order of 2ζ, is a localized modulation of the refractive index, which can reflect a fraction of the probe beam [57]. The measurement of the velocity obviously depends on the knowledge of the film thickness, often obtained by X-ray reflectivity; the uncertainty about thickness is one of the leading terms in the uncertainty to be associated to the final results.

Another factor affecting the precision of results is the resolution by which the arrival times of the pulses are detected. The modulation of the reflectivity has a finite width, which is determined by the space width of the strain pulse. In turn, this space width is determined by the absorption length ζ. Aluminum has one of the shortest absorption lengths, at least for the 800 nm wavelength, other metals, like copper, have longer absorption lengths. In order to limit the absorption length, and thus increase the resolution of the measurement, the deposition of a thin aluminum layer (tens of nanometers) on the sample is a common practice [58]. If this interaction layer is adopted, its presence cannot be neglected in the analysis of results: the additional layer contributes to the vibrational behavior of the structure being investigated. However, operation also in semiconductors, in which the absorption length is significantly longer, has been demonstrated [59].

In several cases, for film thicknesses below $\sim$100 nm, resolution of single echos turns out to be difficult, or impossible [60–62]. However, by a detailed analysis of the reflectivity signal, and taking into account features like the Brillouin oscillations, it has been possible to measure Pt and Fe films of thickness down to 5 nm, deposited on Si or on borosilicate glass substrates [60], and a buried TaN layer of thickness of 20 nm [58]. Layers down to very few nanometers, stacked in a periodic Mo/Si superlattice, were also investigated, exploring different periodicities within a same total thickness. The superposition of two different signals was found. One signal corresponds at the multilayer which act as a single homogeneous effective layer. The other signal corresponds to the multilayer which acts as a Bragg reflector, and confines, in the neighborhood of the outer surface, a mode, which has been called 'localized surface mode'. This mode is sensitive to fine details of the superlattice structure, namely on the outermost layer being the one with the higher or the lower acoustic impedance (in this case Si), and on the presence of the native, nanometric, oxide layer on the Si surface. These details are consistent with the X-ray reflectivity measurements, and allowed to correctly predict the acoustic behavior [63].

The picoseconds ultrasonics technique applied to laterally homogeneous specimens measures the elastic constants involved in the propagation of plane waves traveling perpendicular to the surface: only the out-of-plane elastic characterization of the film is achieved. To overcome this limitation, non homogeneous interaction layers interaction layers have been exploited, namely cut by lithographic techniques to obtain periodic structures [64–66]. It was thus possible to excite vibrational modes of different types [64]: modes of single specific structures, either nanopillars [64, 65] or nanowires [66], collective modes of these nanostructures, coupled via their substrate, and modes of the substrate layer, traveling not only in the direction perpendicular to the surface. The deposition of a metallic grating on a transparent sample allowed to diffract the pump pulse in different orders, obtaining what has been called time-domain Brillouin scattering, and measuring a whole range of acoustic frequencies in a single optical configuration [67].

The acousto-optic (or photoelastic) coupling mechanism was elucidated long ago for bulk samples and for supported films. In the case of free standing films, or membranes, it is increasingly understood that the geometric modulation of the external surfaces by the acoustic modes has a significant role. The denomination of 'moving interface effect' has been proposed [68], which can be seen as a generalization of the ripple effect, which active at the surface of a semi-infinite medium, either homogeneous or layered.

Detailed analyses have been performed for free standing membranes [8], experimentally confirmed with non-constrained, single crystal, Si nanomembranes (thickness 260 nm), for nanoscale structures like cavities and waveguides [68], and for integrated photonic waveguides in on-chip systems [69].

With a sub-wavelength confinement, surface effects play a significant role, and the confinement induced modifications must be taken into account for both the electromagnetic and the acoustic fields. The interaction between these fields turns out to be orders of magnitude more intense than in bulky samples. In volumetric samples the laser beam is only a probe that senses the thermally excited vibrational states, while in strongly confined media the 'stimulated Brillouin scattering' is achievable, in which the electromagnetic beam excites some acoustic mode, and then interacts with it. The strong energy exchange between the trapped light and the acoustic modes, for which the name 'Brillouin optomechanics' has been proposed [70], is exploited in a whole new breed of chip based devices, which include lasers, amplifiers, filters, delay lines and isolators [70, 71]. Here, we do not address this rapidly growing 'optomechanics' field, since we are here focused on the measurement techniques, rather than the device development.

The picosecond ultrasonics technique has proven to be useful also in the measurements of nanorods. In the case of nanorods, the effects of confinement manifest themselves to a high degree. The dispersion relations of both photons and phonons are modified in the nanowires, enhancing interactions and generating peculiar phenomena like stimulated Brillouin scattering, induced transparency, 'slow light' and 'fast light' [72]. The concept of parallel wavevector $\boldsymbol{k}_{\parallel}$ remains meaningful, but can only have the fixed direction of the nanorod, meaning that scattering can only be forward, with of $\boldsymbol{k}_{\parallel} = 0$, i.e. probing resonant phononic modes, or backward, with of $\boldsymbol{k}_{\parallel} = 2\boldsymbol{q}_{\parallel}$ ($\boldsymbol{q}_{\parallel}$ being the component of the optical wavevector), probing traveling phonons. Since, furthermore, the focusing spot is wider than the nanorod, illumination is homogeneous across the nanorod.

Some experiments have been on single suspended nanowires. Copper nanorods (diameter 200 nm, length up 5 μm) suspended across a lithographically obtained trench have been teste by standard techniques for picosecond ultrasonics (100 fs pulses, at the wavelength of 800 nm) [73]. The breathing modes at $\boldsymbol{k}_{\|} = 0$ have been detected.

However, most experiments have been performed on 'forests' of nanowires, either grown on particle seeds, regularly or irregularly positioned, or obtained by etching techniques. In the case of regular positions they form a photonic crystal, in which electromagnetic modes specific of single nanowires, and therefore sensitive to the nanowire diameter, can be detected, but also, depending on the period, also collective electromagnetic modes, sensitive to the photonic crystal period, can be observed. Nanorods with a diameter of the order of 100 nm, heights of several hundred nanometers, and periods ranging from few hundred nanometers to few micrometers have been produced and investigated.

One of the first observation of the vibrational modes of nanorods was obtained by the standard set up for cw Brillouin spectroscopy. Scattering was observed from monocrystalline GaN nanowires, of wurtzite structure, several micrometers long, spontaneously nucleated at the irregular surface of a GaN matrix layer over a Si substrate. Measurements were compared to a finite elements simulation; the analysis had to take into account the dispersion of the diameters, with an average value of 190 nm and a variance of 40 nm. Several branches of the dispersion relation could however be identified [74].

Successive investigations were performed by the picosecond ultrasonics technique, with focused beam spots of tens of micrometers, illuminating a high number of nanorods. GaAs nanorods were fabricated by a standard etching process from a pristine GaAs substrate, covered by a thin gold layer, such that each nanorod had at its tip a gold disk, which acted as transducer, absorbing the pump pulse and launching the acoustic pulse along the nanorod [75].

A square lattice was produced, with nanorods 720 nm long, exploring intervals of diameter (130–270 nm) and of period (300–350 nm). Square lattices of InP nanowires, grown over gold seed particles in regular pattern, with 180 nm diameter and period of 400 nm, were measured by a pump and probe technique. Measurements were compared to a detailed computation of the complicated dispersion relation [76]. This measurement allowed to validate the whole technique, which was then applied to GaAs nanorods, grown over gold seeds in an irregular pattern. Nanorods with the normal zincblende structure, as well as with the wurtzite structure, which in the bulk form is metastable, and was not previously measured, have been investigated; Diameters respectively of 100 nm and 70 nm, with lengths above 1.5 μm. The elastic constants were derived with a resolution sufficient to appreciate the difference between the two structures [77]. Hexagonal lattices of hexagonal GaAs pillars were also produced, of diameters between 103 and 135 nm, exploring lengths up to several micrometers, and pitches from 700 nm to 3 μm [78]. Also this measurement was accompanied by a detailed modeling, allowing to detect up to 10 branches.

The picosecond ultrasonics technique can also investigate nanoobjects, whose aspect ratio is not too far from unity, and in which confinement occurs at its maximum degree. The same concept of parallel wavevector $\boldsymbol{k}_{\|}$ loses its significance, only standing waves exist, whose configuration strongly dependes on the geometry of the object. An extensive and detailed review of work in this area was given [79], and a more recent one focused on the measurement techniques [80]. Recent works include the analysis of compound nanoparticles [81], analyses concerning the exploitation of nanoplates as antennas in the conversion from the laser pulse to the acoustic pulse [82], and work in the direction of the imaging of acoustic modes [83].

4.3 Transient grating spectroscopy

The Transient grating spectroscopy (TGS) technique, which is also called Impulse Stimulated Thermal Scattering (ISTS) or Impulsive Stimulated Scattering (ISS) is an evolution of the Laser ultrasonics technique. As its 'parent' technique, it is appropriate for samples having an external planar surface of sufficient width, and is perfectly suited for the measurement of films. In the conventional laser ultrasonics technique the excitation is impulsive in time, and localized in space (the laser pulse is typically focused into a line by a cylindrical lens), resulting in a broadband pulse which is launched, and detected at some distance. The transit time is thus measured. The possibility of excitation over a more extended region, with a periodic pattern, resulting in a narrowband pulse, was considered in order to increase the efficiency in the generation of the strain pulse. A periodic pattern, of periodicity $\lambda_{\|}$, can be obtained either by two beams which interfere at the surface of the sample [84] or by a hyperbolic diffraction grating [85]. The strain field generated by the pulse is the superposition of a broad, non-oscillating feature, due to the overall heating, whose decay is governed by heat diffusion, and an oscillating feature [74, 86, 87]. The oscillating signal is due to the SAWs which are launched in opposite directions, and which form a standing wave, of periodicity $\lambda_{\|}$, which is eventually damped.

The 'pump' pulse can range from the ns to the ps; wavelengths around 500 nm are typically adopted, but both longer (1064 nm) or shorter wavelengths have been adopted. The strain transient is measured by a probe laser, either a cw laser, or a pulsed laser, in a pump-and-probe scheme. The temperature relaxation is observed by the modulation of the reflectivity of the surface. The surface corrugation due to SAWs can be detected by either the deflection of the strongly focused probe laser [87], or superposing the reflected probe beam with a reference beam, in a heterodyne amplification scheme.

A boxcar configuration has become the standard one for this technique: a volumetric diffraction grating, a 'phase mask' splits the pump pulse into diffraction orders, the +1 and − 1 orders are isolated and overlapped at the sample surface, using a 4*f* imaging system. In this configuration the heterodyne detection is adopted, with the pump and the probe beams which share the same optics, achieving a good phase stability [88–92]. While in principle the explored periodicities $\lambda_{\|}$ can vary over a very interval, practical limitations limit this interval. Firstly, the focusing spots of the lasers, which cannot be too small to avoid power densities that would damage the sample, also cannot be too wide, to avoid low power densities which would lead to too weak signals. Since the focusing spot must contain at least a certain number of interference fringes, to well define the periodicity, this set an upper limit to the accessible periodicities $\lambda_{\|}$. Lower limits are set by the velocity of the detection electronics, typically avalanche photodiodes: (shorter wavelengths mean higher frequencies) and by an intrinsic limitation: the amplitude of the surface displacement, therefore its detectability, scales as the periodicity $\lambda_{\|}$. The practical lower limit of $\lambda_{\|}$ is 2–3 μm for metals, and 5–8 μm for ceramics [6].

The experimental techniques are undergoing further developments. Among them, further refinements of the heterodyne detection [93, 94] and a technique which allows a continuous tuning of the periodicity $\lambda_{\|}$ [95]. Among recent developments, the coherent space and time control of the relative phases of the interference patterns generated by successive laser pulses, which enables the control the periodic surface deformation induced by the pump pulses, which is also monitored by time-resolved X-ray reflectivity [96], and the exploitation of femtosecond pulses of extreme ultraviolet light (12.7 nm) to obtain surface periodicities $\lambda_{\|}$ of 280 nm [97]. Furthermore, the possibility of performing large area 2D maps, also dealing with surface roughness [98].

Among recent applications of the technique, the characterization of nanocrystalline diamond coatings [99], and the exploitation of the optical character of the technique to perform measurements at different temperatures [100], and in situ measurements [101]. As previously mentioned, the possibility of tuning the periodicity $\lambda_{\|}$ allows to choose the depth which is probed. This is particularly useful in ion irradiation experiments: the ion implantation depth is of the order of micrometers, meaning that the irradiation modified skin is of that order; the periodicity $\lambda_{\|}$ can be tuned in order to investigate precisely the depth affected by irradiation [6, 102–104].

4.4 Brillouin spectroscopy

Brillouin spectroscopy investigates the inelastic scattering of light by vibrational excitations of the acoustic type [105, 106]. It is therefore in principle able to investigate both traveling waves and standing waves, from bulky samples to films and to nanorods, and the name 'Brillouin optomechanics' has been proposed for the interaction between the electromagnetic field and the acoustic field in nanoobjects. The ability of Brillouin spectroscopy to single out a wavevector, $\boldsymbol{k}$ or $\boldsymbol{k}_{\|}$ makes it attractive whenever a wavevector can meaningfully be considered. As a tool for the measurement of material properties, surface Brillouin scattering (Brillouin scattering by SAWs) has been extensively used to characterize supported films [107–109].

Brillouin spectroscopy is performed illuminating the sample by a focused laser beam, of wavevector $\boldsymbol{q}_i$ and circular frequency $\boldsymbol{\Omega}_i$, and collecting the light scattered along a direction $\boldsymbol{q}_s$. The spectral content of the scattered light is dominated by the light elastically scattered, at $\boldsymbol{\Omega}_i$, but can also contain the Stokes/anti-Stokes doublets due to inelastic scattering by thermally excited vibrations of circular frequency ω, at frequencies $\boldsymbol{\Omega}_s = \boldsymbol{\Omega}_i \pm \omega$. Since the scattering geometry selects the probed wavevectors $\boldsymbol{k} = \pm(\boldsymbol{q}_s - \boldsymbol{q}_i)$, or, in the case of SAWs, $\boldsymbol{k}_{\|} = \pm(\boldsymbol{q}_s - \boldsymbol{q}_i)_{\|}$, from each spectrum one point of the dispersion relation $\omega = \omega(\boldsymbol{k})$, or $\omega = \omega(\boldsymbol{k}_{\|})$, is obtained for each of the branches for which a spectral peak is measured [40, 105–108].

Scattering by bulk waves occurs by the elasto-optic effect, the modulation of the refractive index by the strain: a periodic modulation, of periodicity represented by $\boldsymbol{k}$, is a weak diffraction grating, in motion at the speed of sound. Obviously scattering by bulk waves only occurs in sufficiently transparent media, in which refraction must be taken into account: the wavevectors $\boldsymbol{q}_i$ and $\boldsymbol{q}_s$ are refracted into $\boldsymbol{q'}_i$ and $\boldsymbol{q'}_s$, and the probed wavevector is more precisely $\boldsymbol{k} = \boldsymbol{q'}_s - \boldsymbol{q'}_i$. The refractive index n has therefore a role. Scattering by surface waves, whose strain field penetrates in an outermost layer, occurs by the same mechanism, if the medium is sufficiently transparent, and by the ripple effect: the periodic corrugation of the surface induced by the wave, which also is a weak diffraction grating. In this case the probed wavevector $\boldsymbol{k}_{\|}$ does not depend on the refractive index, since Snell's refraction law states that, for any optical wavevector $\boldsymbol{q}$, the component $\boldsymbol{q}_{\|}$ parallel to the surface remains unchanged upon refraction. In metals scattering occurs only by surface waves, by the ripple effect.

The scattering process can also be described as follows. The spontaneous, chaotic, thermal motion can be thought as being three-dimensionally Fourier transformed into an incoherent superposition of harmonic waves, whose wavevectors have all the possible values. The scattering process probes the component of this decomposition which has precisely the wavevector $\boldsymbol{k}$, or $\boldsymbol{k}_{\|}$, selected by the scattering geometry. The exploitation of the thermal motion means that the excitation has the broadest band, but has small amplitude, implying, in many cases, time consuming measurements.

Brillouin spectroscopy has been used to characterize bulk materials, and surface Brillouin spectroscopy is particularly suited for the characterization of thin films, which are of our concern here. Although Brillouin spectroscopy can, in principle, be performed in various geometries [105, 108, 109], possibly scanning a wide interval of $|\boldsymbol{k}|$, or of $|\boldsymbol{k}_{\|}|$, the backscattering configuration ($\boldsymbol{q}_s = -\boldsymbol{q}_i$) is, by far, the most frequently adopted. This occurs because the backscattering geometry maximizes $|\boldsymbol{k}|$, and its implementation is more practical. In backscattering the probed wavevector is $\boldsymbol{k} = \pm 2\boldsymbol{q}'_i$, with $|\boldsymbol{q}'_i| = n|\boldsymbol{q}_i| = n2\pi/\lambda_{opt}$, where λ_{opt} is the optical wavelength of the incident beam (remember that n is simply a scalar only when symmetry is at least cubic), or $\boldsymbol{k}_{\|} = \pm 2(\boldsymbol{q}_i)_{\|}$, with $\left|(\boldsymbol{q}_i)_{\|}\right| = |\boldsymbol{q}_i| \sin\theta$, where θ is the incidence angle (the angle between the incident beam and the normal to the surface). The wavevector $\boldsymbol{k}_{\|}$ does not depend on n, and depends on the incidence angle. Since incidence cannot be close to the normal (to avoid the specular reflections) nor too close to grazing one (because of the decline of the scattering cross section) the practically accessible range $(\sin\theta)_{max}/(\sin\theta)_{min}$ seldom exceeds the value of 2.

For SAWs, the above relations give $\lambda = \lambda_{opt}/(2n)$ and $\lambda_{\|} = \lambda_{opt}/(2\sin\theta)$. This means that, with the often adopted λ_{opt} =532 nm, for supported films of thickness D of a couple of micrometers, or more, the condition $\lambda_{\|} \ll D$ is easily accessible. We remember that in this case the Rayleigh wave at the surface of the film is insensitive to the substrate properties and to the precise value of thickness, and gives a direct access to the film properties. When films of this thickness are sufficiently transparent, also scattering by bulk waves is observable, the bulk waves being fully developed. Typical properties of metals, semiconductors and ceramics give, for these wavelengths, frequencies ranging from few GHz to several tens of GHz.

Brillouin spectroscopy was extensively exploited to characterize tetrahedral amorphous carbon films of thicknesses of hundreds of nanometers [110], tens of nanometers [111], down to a few nanometers [112]. In the case of thicker tetrahedral amorphous carbon films [3 micrometers] a detailed characterization was achieved by combining Brillouin spectroscopy and laser ultrasonics. The combination of the techniques gave access to a wide range of frequencies, allowing detailed determination of the elastic properties of the film [113]. Brillouin spectroscopy turned out to be a useful characterization tool also for other types of films of interest in materials science, like boron films [114] and amorphous and nanocrystalline tungsten films [115].

Since picosecond ultrasonics characterizes the out-of-plane properties by waves traveling normal to the surface, while Brillouin spectroscopy characterizes the in-plane properties by waves traveling along the surface, the two techniques have also been exploited in a combined way, achieving a more complete characterization [62, 116–118].

Brillouin spectroscopy also lends itself to the characterization of structures other than films or layers. In particular, single-walled carbon nanotubes were characterized, measuring Brillouin scattering by a free-standing film of pure, partially aligned, single-walled nanotubes, and analyzing the results in terms of continuum models [119]. The dependence of the measured spectra on the angle between the exchanged wavevector and the preferential direction of the tubes shows that the tube-tube interactions are weak: the tubes are vibrationally almost independent. The tubes are modeled as continuous membranes; taking into account that AFM images suggest that the tube segments contributing to scattering are not in the infinite tube length approximation, it was possible to derive the 2D Young modulus for the tube wall, achieving the first dynamic estimation of the stiffness of the tube wall. Scattering from carbon nanotubes was observed also in a different geometry, with an ordered array of tubes, clamped at one end [120]. Brillouin spectroscopy also allowed one of the first observations of the vibrational modes of nanorods [74].

5. Conclusions

Nanotechnology, and nanodevices, identify a rapidly growing technological field. For the development of nanodevices a precise knowledge of the elastic properties of materials is of utmost importance, also because materials obtained at the nanoscale have properties which do not coincide with those of their bulk counterpart. Accordingly, a variety of techniques have been developed, which have proven able to investigate the material properties at the nanoscale. Most of these techniques rely on the interaction between optical electromagnetic fields, and mechanical acoustic fields. An overview has been given of these various techniques, offering elements for the evaluation of their appropriateness for different characterization needs.

References

[1] Auld BA. Acoustic fields and Waves in Solids. Malabar, Florida: Robert E. Krieger Publishing Company; 1990. ISBN: 0898747821

[2] Every AG. The Elastic Properties of Solids: Static and Dynamic Principles, In: Levy M, Bass H, Stern R, Keppens V, editors. Handbook of Elastic Properties of Solids, Liquids, and Gases; Volume I: Dynamic Methods for Measuring the Elastic Properties of Solids. New York: Academic Press; 2001. p. 3–36. ISBN: 01–244–5760-6

[3] Kundu, T. Mechanics of elastic waves and ultrasonics non-destructive evaluation. In Kundu, T. editor. Ultrasonic and Electromagnetic NDE for Structure and Material Characterization: Engineering and Biomedical Applications. Boca Raton, FL: CRC Press; 2012. p. 1–98. ISBN: 9781439836637.

[4] Levy M, Bass H, Stern R, Keppens V, editors. Handbook of Elastic Properties of Solids, Liquids, and Gases; Volume I: Dynamic Methods for Measuring the Elastic Properties of Solids. New York: Academic Press; 2001. ISBN: 01–244–5760-6

[5] Farnell W, Adler EL (1972). Elastic wave propagation in thin layers. In: Mason WP, Thurston RN, editors. Physical Acoustics, Vol. 9. New York: Academic; 1972. p. 35–127.

[6] Hofmann F, Short MP, Dennett CA. Transient grating spectroscopy: An ultrarapid, nondestructive materials evaluation technique. MRS Bulletin. 2019;44:392–402. DOI: 10.1557/mrs.2019.104.

[7] Guyader J-L. Vibration in continuous media. Wiley; 2013. DOI: 10.1002/9780470612453.

[8] Graczykowski B, Gueddida A, Djafari-Rouhani B, Butt H-J, Fytas G. Brillouin light scattering under one-dimensional confinement: Symmetry and interference self-canceling. Physical Review B. 2019;99:165431. DOI: 11.1103/PhysRevB.99.165431.

[9] Brekhovskikh LM. Waves in layered media, 2nd Edition. London: Academic Press; 1980. ISBN: 0–12–130560-0.

[10] Brekhovskikh LM, Godin O. Acoustics of Layered media I. Berlin: Springer-Verlag; 1990. DOI: 10.1007/978-3-642-52369-4.

[11] Royer D, Dieulesaint E. Elastic waves in solids I: free and guided propagation. Advanced texts in physics. Berlin–Heidelberg: Springer, 1999. ISBN: 978–3–540-65932-7.

[12] Garcia-Sanchez D, Déleglise S, Thomas J-L, Atkinson P, Lagoin C, Perrin B. Acoustic confinement in superlattice cavities. Phys Rev A. 2016; 94:033813. DOI: 10.1103/PhysRevA.94.033813.

[13] Visscher WM, Migliori A, T.Bell M, Reinert RA. On the normal modes of free vibration of inhomogeneous and anisotropic elastic objects. J Acoust Soc Am. 1991;90:2154–2162. DOI: 10.1121/1.401643.

[14] Mante P-A, Belliard L, Perrin B. Acoustic phonons in nanowires probed by ultrafast pump-probe spectroscopy. Nanophotonics. 2018;7:1759–1780. DOI: 10.1515/nanoph-2018-0069.

[15] Nishiguchi N, Ando Y, Wybourne MN. Acoustic phonon modes of rectangular quantum wires. J Phys: Condens. Matter 1997;9:5751–5764. DOI: 10.1088/0953-8984/9/27/007.

[16] Li G, Lamberton Jr GA, Gladden JR. Acoustic modes of finite length homogeneous and layered cylindrical shells: Single and multiwall carbon

nanotubes. J Appl Phys. 2008;104: 033524. DOI: 10.1063/1.2965187.

[17] Martínez-Gutiérrez D, Velasco VR. Acoustic waves of GaN nitride nanowires. Surf Sci. 2011;605:24–31. DOI: 10.1016/j.susc.2019.09.017.

[18] Mizuno S, Nishiguchi N. Acoustic phonon modes and dispersion relations of nanowire superlattices. J Phys: Condens Matter. 2009;21:195303. DOI: 10.1088/0953-8984/21/19/195303.

[19] Jean C, Belliard L, Becerra L, Perrin B. Backward propagating acoustic waves in single gold nanobeams. Appl Phys Letters. 2015;107:193103. DOI: 10.1063/1.4935351.

[20] Finot E, Passian A, Thundat T. Measurement of mechanical properties of cantilever shaped materials. Sensors. 2008;8:3497–3541. DOI: 10.3390/s8053497.

[21] Hasheminejad SM, Gheshlaghi B. Eigenfrequencies and quality factors of nanofilm resonators with dissipative surface stress effects. Wave motion. 2013;50:94–100. DOI: 10.1016/j.wavemoti.2012.07.001.

[22] Chaste J, Eichler A, Moser J, Ceballos G, Rurali R, Bachtold A. A nanomechanical mass sensor with yoctogram resolution. Nature Nanotechnol. 2012;7:301–304. DOI: 10.1038/NNANO.2012.42.

[23] Schmid S, Villanueva LG, Roukes ML. Fundamentals of nanomechanical resonators. Cham, Switzerland: Springer; 2016. DOI: 10.1007/978-3-319-28689-1.

[24] Zhou J, Moldovan N, Stan L, Cai H, Czaplewski DA, Lopez D. Approaching the strain-free limit in ultrathin nanomechanical resonators. Nano Letters. 2020;20:5693–5698. DOI: 10.1021/acs.nanolett.0c01027.

[25] Czaplewski DA, Sullivan JP, Friedmann TA, Wendt JR. Mechanical dissipation in tetrahedral amorphous carbon. J Appl Phys. 2005;97:023517. DOI: 10.1063/1.1821638.

[26] Czaplewski DA, Sullivan JP, Friedmann TA, Wendt JR. Temperature dependence of the mechanical properties of tetrahedrally coordinated amorphous carbon thin films. Applied Physics Letters. 2005;87:161915. DOI: 10.1063/1.2108132.

[27] Hassan MO, Nojeh A, Takahata K. A microcantilever of self-suspended carbon nanotube forest for material characterization and sensing applications. Appl Phys Lett. 2020;117: 013101. DOI: 10.1063/5.0012533.

[28] Vallabhaneni AK, Rhoads JF, Murthy JY, Ruan X. Observation of nonclassical scaling laws in the quality factors of cantilevered carbon nanotube resonators. J Appl Phys. 2011;110: 034312. DOI: 10.1063/1.3611396.

[29] Knoebber F, Zuerbig V, Heidrich N, Hees J, Sah RE, Baeumler M, Leopold S, Paetz D, Ambacher O, Lebedev V. Static and dynamic characterization of AlN and nanocrystalline diamond membranes. Thin Solid Films. 2014;558: 267:271. DOI: 10.1016/j.tsf.2014.03.015.

[30] Hees J, Heidrich N, Pletschen W, Sah RE, Wolfer M, Williams OA, Lebedev V L, Nebel CE, Ambacher O. Piezoelectric actuated micro-resonators based on the growth of diamond on aluminum nitride thin films. Nanotechnology. 2013;24:025601. DOI: 10.1088/0957-4484/24/2/025601.

[31] Zuerbig V, Hees J, Pletschen W, Sah RE, Wolfer M, Kirste L, Heidrich N, Nebel CE, Ambacher O, Lebedev V. Elastic properties of of ultrathin diamond/AlN membranes. Thin Solid Films. 2014;558:267:271. DOI: 10.1016/j.tsf.2014.03.015.

[32] Castellanos-Gomez A, van Leeuven R, Buscema M, van der Grant HSJ, Steele GA, Venstra WJ. Single-Layer MoS2 Mechanical Resonators. Advanced Materials. 2013;25:6719–6723. DOI: 10.1002/adma.201303569.

[33] Kim JY, Chung HJ, Kim HJ, Cho HM, Yang HK, Park JC. Surface acoustic wave propagation properties of nitrogenated diamond-like carbon films, J of Vacuum Science and Technology A. 2000;18:1993—1997. DOI: 10.1116/1.582460.

[34] Bi B, Huang W-S, Asmussen J, Bolding B. Surface acoustic waves on nanocrystalline diamond. Diamond and Related Materials. 2002;11:677–680. DOI: 10.1016/S0925-9653(01)00621-5.

[35] Shih W-C, Wang M-J, Nan Lin I. Characteristics of ZnO thin film surface acoustic wave devices fabricated using nanocrystalline diamond film on silicon substrate. Diamond & Related Materials. 2008;17:390–395. DOI: 10.1016/j.diamond.2008.01.082.

[36] D'Evelyn MP, Taniguchi T. Elastic properties of translucent polycrystalline cubic boron nitride as characterized by the dynamic resonance method. Diamond and Related Materials. 1999;8: 1522–1526. DOI: 10.1016/S0925-9635(99)00077-1.

[37] Nieves FJ, Gascòn F, Bayòn A. Precise and direct determination of the elastic constants of a cylinder with a length equal to its diameter. Review of Scientific Instruments. 2000;71:2433–2439. DOI: 10.1063/1.1150632.

[38] ASTM International. E1876–15 Standard Test Method for Dynamic Young's Modulus, Shear Modulus, and Poisson's Ratio by Impulse Excitation of Vibration. West Conshohocken, PA: ASTM International. 2011. DOI: 10.1520/E1876-15.

[39] Ruello P, Gusev VE. Physical mechanisms of coherent acoustic phonon generation by ultrafast laser action. Ultrasonics. 2015;56:21–35. DOI: 10.1016/j.ultras.201406.004

[40] Sandercock JR. Trends in Brillouin scattering: Studies of opaque materials, supported films, and central modes. In: Cardona M, Güntherodt G, editors. Light Scattering in solids III. Topics in Applied Physics, vol 51. Berlin: Springer; 1982. p. 173–206. DOI: 10.1007/3540115137_6

[41] Neubrand A, Hess P. Laser generation and detection of surface acoustic waves: Elastic properties of surface layers. Journal of applied physics. 1992;71:227–238. DOI: 10.1063/1.350747

[42] Whitfield MD, Audic B, Flannery CM, Kehoe LP, Crean GM, Jackman RB. Characterization of acoustic Lamb wave propagation in polycrystalline diamond film by laser ultrasonics. Journal of Applied Physics. 2000;88:2984–2993. DOI: 10.1063/1.1286010

[43] Lehmann G, Hess P, Weissmantel S, Reisse G, Scheible P, Lunk, A. Young's modulus and density of nanocrystalline cubic boron nitride films determined by dispersion of surface acoustic waves. Applied Physics A. 2002;74:41–45. DOI: 10.1007/s003390100897

[44] Schneider D, Schwarz T, Scheibe H-J, Panzner M. Non destructive evaluation of diamond-like carbon films by laser induced surface acoustic waves. Thin Solid Films. 1997;295:107–116. DOI: 10.1016/S0040-6090(96)09163-8

[45] Schneider D, Witke T, Schwarz T, Schoneich B, Schultrich B. Testing ultra-thin films by laser-acoustics. Surface and Coatings Technology. 2000;126: 136–141. DOI: 10.1016/S0257-8972(99)00672-6

[46] Coufal H, Grygier R. Broadband detection of laser-excited surface acoustic waves by a novel transducer

employing ferroelectric polymers. J Acoustical Society of America. 1992;92: 2980. DOI: 10.1121/1.404363

[47] Lee YC, Lin SJ, Buck V, Kunze R, Schmidt H, Lin CY, Fang WL, Lin IN. Surface acoustic wave properties of natural smooth ultra-nanocrystalline diamond characterized by laser-induced SAW pulse technique. Diamond & Related Materials. 2008;17:446–450. DOI: 10.1016/j.diamond.2007.08.25

[48] Webersen M, Johannesmann S, Duechting J, Claes L, Henning B. Guided ultrasonic waves for determining effective orthotropic material parameters of continuous fiber reinforced thermoplastic plates. Ultrasonics. 2018,84:53–62. DOI: 10.1016/j.ultras.2017.10.005

[49] Yang F, Dorantes-Gonzalez DJ, Chen K, Lu Z, Jin B, Li Y, Chen Z, Hu X. An integrated laser-induced piezoelectric/differential confocal surface acoustic wave system for measurement of thin filmYoung modulus. Sensors. 2012;12:12208–12219. DOI: 10.3390/s120912208

[50] Zhang Q, Xiao X, Cheng Y-T, Verbrugge MW. A non destructive method for measuring the mechanical properties of ultrathin films prepared by atomic layer deposition. Applied Physics Letters. 2014;105:061901. DOI: 10.1063/1.4892539

[51] Thomsen C, Strait J, Vardeny Z, Maris HJ, Tauc J, Hauser JJ. Coherent phonon generation and detection by picoseconds light pulses. Physical Review Letters. 1984;53:989–992. DOI: 10.1103/PhysRevLett.53.989

[52] Thomsen C, Grahn HT, Maris HJ, Tauc J. Surface generation and detection of phonons by picoseconds light pulses. Physical Review B. 1986;34:4129–4138. DOI: 10.1103/PhysRevB.34.4129

[53] Wright OB, Kawashima K. Coherent phonon detection from ultrafast surface vibrations. Phys. Rev. Letters. 1992;69: 1668–1671. DOI: 10.1103/PhysRevLett.69.1668.

[54] Wright OB. Thickness and sound velocity measurement in thin transparent films with laser picosecond acoustics. J. Appl. Phys. 1992;71:1617–1629. DOI: 10.1063/1.351218.

[55] Devos A, Côte R. Strong oscillations detected by picoseconds ultrasonics in silicon: evidence for an electronic structure effect. Physical Review B. 2004;70:125208. DOI: 10.1103/PhysRevB.70.125208

[56] Vollmann J, Profunser DM, Dual J. Sensitivity improvement of a pump-probe set-up for thin film and microstructure metrology. Ultrasonics. 2002;40:757–763. DOI: 10.1016/S0041-624X(02)00207-X

[57] Yu K, Devkota T, Beane G, Wang GP, Hartland GV. Brillouin oscillations from single Au nanoplate opto-acoustic transducers. ACS Nano. 2017;11:8064–8071. DOI: 10.1021/acsnano.7b2703

[58] Bryner J, Profunser DM, Vollmann J, Mueller E, Dual J. Characterization of Ta and TaN diffusion barriers beneath Cu layers using picosecond ultrasonics. Ultrasonics. 2006;44:e1269-e1275. DOI: 10.1016/j.ultras.2006.05097

[59] Babilotte P, Ruello P, Mounier D, Pezeril T, Vaudel G, Edely M, Breteau J-M, Gusev V. Famtosecond laser generation and detection of high-frequency acoustic phonons in GaAs semiconductors. Physical Review B. 2010;81:245207. DOI: 10.1103/PhysRevB.81.245207

[60] Ogi H, Fujii M, Nakamura N, Shagawa T, Hirao M. (2007). Resonance acoustic-phonon spectroscopy for studying elasticity of ultrathin films. Applied Physics Letters. 2007;90: 191906. DOI: 10.1063/1.2737819

[61] Daly BC, Bailey ST, Sooryakumar R, King SW. Noncontact optical metrologies for Young's modulus measurements of nanoporous low-k dielectric thin films. Journal of Nanophotonics. 2013;7:073094. DOI: 10.1117/1.JNP:7.07394

[62] Faëse F, Poinot Cherroret D, Chatel S, Becerra L, Challali F, Djemia P, Belliard L. Mechanical properties of elementary layers involved in a multilayer optical stack by photon-acoustic phonon interaction approaches, J. Appl. Phys. 2018;124:125307. DOI: 10.1063/1.5030749

[63] Belliard L, Huynh A, Perrin B, Michel A, Abadias G, Jaouen, C. Elastic properties and phonon generation in Mo/Si superlattices. Physical Review B. 2009;80:155424. DOI: 10.1103/PhysRevB.80.155424.

[64] Bienville T, Robillard JF, Belliard L, Roch_Jeune I, Devos A, Perrin B. (2006). Individual and collective vibrational modes of nanostructures studied by picosecond ultrasonics. Ultrasonics. 2006;44:e1289-e1294. DOI: 10.1016/j.ultras.2006.05.179

[65] Mante PA, Robillard JF, Devos A. Complete thin film mechanical characterization using picoseconds ultrasonics and nanostructured transducers: experimental demonstration on SiO2. Applied Physics Letters. 2008;93:071909. DOI: 10.1063/1.2975171

[66] Bjornsson MM, Connolly AB, Mahat S, Rachmilowitz BE, Daly BC, Antonelli GA, Myers A, Singh KJ, Yoo HJ, King SW. Picosecond ultrasonic study of surface acoustic waves on titanium nitride nanostructures. J. Appl. Phys. 2015;117:095305. DOI: 10.1063/1.4914048

[67] Matsuda O, Pezeril T, Chaban I, Fujita K, Gusev V. Time-domain Brillouin scattering assisted by diffraction gratings. Physical Review B. 2018;97:064301. DOI: 10.1103/PhysRevB.97.064301

[68] Pennec Y, Laude V, Papanikolaou N, Djafari-Rouhani B, Oudich M, El Jallal S, Beugnot JC, Escalante JM, Martínez A. Modeling light-sound interaction in nanoscale cavities and waveguides. Nanophotonics. 2014;3:413–440. DOI: 10.1515/nanoph-2014-0004

[69] Wolff C, Steel MJ, Eggleton BJ, Poulton CG. Stimulated Brillouin scattering in integrated photonic waveguides: Forces, scattering mechanisms, and coupled-mode analysis. Physical Review A. 2015;92:013836. DOI: 10.1103/PhysRevA.92.013836

[70] Wiederhecker GS, Dainese P, Mayer Alegre TP. Brillouin optomechanics in nanophotonic structures. APL_Photonics. 2019;4:071101. DOI: 10.1063/1.5088169

[71] Eggleton_BJ, Poulton CG, Rakich PT, Steel MJ, Bahl G. Brillouin integrated photonics. NaturePhotonics. 2019;13:664–677. DOI: 10.1038/s41566-019-0498-z

[72] Kim_JH, Kuzyk MC, Han K, Wang H, Bahl G. Non-reciprocal Brillouin scattering induced transparency. NaturePhysics. 2015;11:275–280. DOI: 10.1038/NPHYS3236

[73] Jean CJ, Belliard L, Cornelius TW, Thomas O, Toimil-Molares ME, Cassinelli M, Becerra L, Perrin B. Direct observation of Gigahertz coherent guided acoustic phonons in free-standing single copper nanowires. J. Phys. Chem. Lett. 2014;5:4100–4104. DOI: 10.1021/jz502170j

[74] Johnson WL, Kim SA, Geiss R, Flannery CM, Bertness KA, Heyliger PR. Vibrational modes of GaN nanowires in the Gigahertz range. Nanotechnology

2012;23:495709. .DOI: 10.1088/0957-4484/23/49/495709

[75] Yang S-C, Wu Y-C, Mante P-A, Chen C-C, Chen H-P, Chou H-Y, Shih M-H, Sun C-K. Efficient excitation of guided acoustic waves in semiconductor nanorods through external metallic acoustic transducer. Appl. Phys. Letters. 2014;105:243101. DOI: 10.1063/1.4904414

[76] Mante P-A, Anttu N, Zhang W, Wallentin J, Chen I-J, Lehmann S, Heurlin M, Borgström MT, Pistol M-E, Yartsev A. Confinement effects on Brillouin scattering in semiconductor nanowire photonic crystal. Physical Review B. 2016;94:024115. DOI: 10.1103/PhysRevB.94.024115

[77] Mante P-A, Lehmann S, Anttu N, Dick KA,Yartsev A. Nondestructive complete mechanical characterization of zinc blende and wurtzite GaAs nanowires using time-resolved pump-probe spectroscopy. NanoLetters. 2016; 16:4792–4798. DOI: 10.1021/acs.nanolett.6b00786

[78] Kargar F, Debnath B, Kakko J-P, Säynätjoki A, Lipsanen H, NikaDL, Lake RK, Balandin AA. Direct observation of confined acoustic phonon polarization branches in free-standing semiconductor nanowires. Nature Communications. 2016;7:13400. DOI: 10.1038/ncomms13400

[79] Crut A, Maioli P, Del Fatti N, Vallée F. Acoustic vibrations of metal nano-objects: Time-domain investigations. PhysicsReports. 2015;549:1–43. DOI: 10.1016/j.physrep.2014.09.004

[80] Balogun_IEEE19] Balogun O. Optically Detecting Acoustic Oscillations at the Nanoscale Exploring techniques suitable for studying elastic wave propagation. IEEE Nanotechnology Magazine. 2019;13:39–54. DOI: 10.1109/MNANO.2019.2905021

[81] Zhao X, Nie Z, Feng Y, Zhao W., Zhang J, Zhang W, Maioli P, Loh Z-H. TI Ultrafast acoustic vibrations of Au-Ag nanoparticles with varying elongated structures. Phys. Chem. Chem. Phys. 2020;22:22728–22735. DOI: 10.1039/d0cp03260c.

[82] Devkota T, Yu K, Hartland GV. Mass loading effects in the acoustic vibrations of gold nanoplates. Nanoscale. 2019;11:16208–16213. DOI: 10.1039/c9nr05940g

[83] Guillet Y, Abbas A, Ravaine S, Audoin B. Ultrafast microscopy of the vibrational landscape of a single nanoparticle. Applied Physics Letters. 2019;114: 091904. DOI: 10.1063/1.5085157

[84] Nelson KA, Dwayne Miller RJ, Lutz DR, Fayer MD. Optical generation of tunable ultrasonic waves. J. Appl. Phys. 1982;53:1144–1149. DOI: 10.1063/1.329864

[85] Huang J, Krishnaswamy S, Achenbach JD. Laser-generation of narrow-band surface waves. In: Proceedings of the IEEE 1991 Ultrasonics Symposium; 1991; Orlando, FL, USA. New York: IEEE; 1991. vol.1, p. 537–541. DOI: 10.1109/ULTSYM.1991.234221.

[86] Fivez J. The determination of the elastic constants of isotropic solids by means of transient thermal surface gratings. J. Applied Physics. 2016;119: 015301. DOI: 10.1063/1.4939208

[87] Kading OW, Skurk H, Maznev AA, Matthias E., Transient thermal gratings at surfaces for thermal characterization of bulk materials and thin films. Applied Physics A. 1995;61:253–261. DOI: 10.1007/BF015381980

[88] Rogers JA, Fuchs M, Banet MJ, Hanselman JB, Logan R, Nelson KA. Optical system for rapid materials characterization with the transient grating technique: Application to

nondestructive evaluation of thin films used in microelectronics. Applied Physics Letters. 1997;71:225. DOI: 10.1063/1.119506.

[89] Maznev AA, Nelson KA, Rogers JA. Optical heterodyne detection of laser-induced gratings. Opt. Letters. 1998;23: 1319–21. DOI: 10.1364/OL.23.001319

[90] Rogers JA, Maznev AA, Banet MJ, Nelson KA. Optical generation and characterization of acoustic waves in thin films: fundamentals and applications. Annu. Rev. Mater. Sci. 2000;30:117–57. DOI: 10.1146/annurev.matsci.30.1.117

[91] Landa M, Verstraeten B, Sermeus J, Salenbien R, Sedlak P, Seiner H, Glorieux C. Thermomechanical properties of single crystals evaluated by impulsive stimulated thermal scattering technique. Journal of Physics: Conference Series. 2011;278:012023. DOI: 10.1088/1742-6596/278/1/012023.

[92] Verstraeten B, Sermeus J, Salenbien R, Fivez J, Shkerdin G, Glorieux C. Determination of thermoelastic material properties by differential heterodyne detection of impulsive stimulated thermal scattering. Photoacoustics. 2015;3:64–77. DOI: 10.1016/j.pacs.2015.05.001

[93] Dennett CA, Short MP. Time-resolved, dual heterodyne phase collection transient grating spectroscopy. Appl. Phys. Letters. 2017; 110:211106 DOI: 10.1063/1.4983716.

[94] Dennett CA, Short MP. Thermal diffusivity determination using heterodyne phase insensitive transient grating spectroscopy. Journal of Applied Physics. 2018;123:215109 DOI: 10.1063/1.5026429

[95] Vega-Flick A, Eliason JK, Maznev AA, Khanolkar A, Abi Ghanem M, Boechler N, Alvarado-Gil JJ, Nelson KA. Laser-induced transient grating setup with continuously tunable period. Rev. Sci. Instrum. 2015;86:123101 DOI: 10.1063/1.4936767

[96] Pudell JE, Sander M, Bauer R, Bargheer M, Herzog M, Gaal P. Full spatiotemporal control of laser-excited periodic surface deformations. Physical Review Applied. 2019;12:024036. DOI: 10.1103/PhysRevApplied.12.024036

[97] Maznev AA, Bencivenga F, Cannizzo A, Capotondi F, Cucini R, Duncan RA, Feurer T, Frazer TD, Foglia L, Frey H-M, Kapteyn H, Knobloch J, Knopp G, Masciovecchio C, Mincigrucci R, Monaco G, Murnane M, Nikolov I, Pedersoli E, Simoncig A, Vega-Flick A, Nelson KA. Generation of coherent phonons by coherent extreme ultraviolet radiation in a transient grating experiment. Applied Physics Letters. 2018;113:221905. DOI: 10.1063/1.5048023.

[98] Reza A, Dennett CA, Short MP, Waite J, Zayachuk Y, Magazzeni CM, Hills S, Hofmann F. Non-contact, non-destructive mapping of thermal diffusivity and surface acoustic wave speed using transient grating spectroscopy. Rev. Sci. Instrum. 2020; 91:054902 DOI: 10.1063/5.0003742.

[99] Sermeus J, Verstraeten B, Salenbien R, Pobedinskas P, Haenen K, Glorieux C. Determination of elastic and thermal properties of a thin nanocrystalline diamond coating using all-optical methods. Thin Solid Films. 2015;590: 284–292. DOI: 10.1016/j.tsf.2015.08.007.

[100] Heczko O, Seiner H, Stoklasova P, Sedlak P, Sermeus J, Glorieux C, Backen A, Fahler S, Landa M. Temperature dependence of elastic properties in austenite and martensite of Ni-Mn-Ga epitaxial films. Acta Materialia. 2018; 145:298–305. DOI: 10.1016/j.actamat.2017.12.011.

[101] Grabec T, SedlákP, Stoklasová P, Thomasová M, Shilo D, Kabla M, Seiner

H, Michal Landa M. In situ characterization of local elastic properties of thin shape memory films by surface acoustic waves. Smart Mater. Struct. 2016;25:127002. DOI: 10.1088/0964-1726/25/12/127002

[102] Hofmann F, Mason DR, Eliason JK, Maznev AA, Nelson KA, Dudarev SL. Non-contact measurement of thermal diffusivity in ion-implanted nuclear materials. Scientific Reports. 2015;5: 16042. DOI: 10.1038/srep16042

[103] Dennett CA, So KP, Kushima A, Buller DL, Hattar K, Short MP. Detecting self-ion irradiation-induced void swelling in pure copper using transient grating spectroscopy. Acta Materialia. 2018;145:496–503. DOI: 10.1016/j.actamat.2017.12.007

[104] Reza A, Zayachuk Y, Yu HB, Hofmann F. Transient grating spectroscopy of thermal diffusivity degradation in deuterium implanted tungsten. Scripta Materialia. 2020;174: 6–10. DOI: 10.1016/j.scriptamat.2019.08.01

[105] Grimsditch M. Brillouin scattering. In: Levy M, Bass H, Stern R, Keppens V, editors. Handbook of Elastic Properties of Solids, Liquids, and Gases; Volume I: Dynamic Methods for Measuring the Elastic Properties of Solids. New York: Academic Press; 2001. p. 331–347. ISBN: 01–244–5760-6.

[106] Comins JD. Surface Brillouin scattering. In: Levy M, Bass H, Stern R, Keppens V, editors. Handbook of Elastic Properties of Solids, Liquids, and Gases; Volume I: Dynamic Methods for Measuring the Elastic Properties of Solids. New York: Academic Press; 2001. p. 349–378. ISBN: 01–244–5760-6.

[107] Every AG. Measurement of the near surface elastic properties of solids and thin supported films. Measurement Science and Technology. 2002;13:R21-R39. DOI: 10.1088/0957-0233/13/5/201

[108] Beghi MG, Every AG, Prakapenka V, Zinin PV. Measurement of the Elastic Properties of Solids by Brillouin Spectroscopy. In Kundu T, editor. Ultrasonic and Electromagnetic NDE for Structure and Material Characterization: Engineering and Biomedical Applications. Boca Raton, FL: CRC Press; 2012. p. 539–610. ISBN: 9781439836637

[109] Beghi MG, Di Fonzo F, Pietralunga S, Ubaldi C, Bottani CE. Precision and accuracy in film stiffness measurement by Brillouin spectroscopy. Review of Scientific Instruments. 2011;82:053107. DOI: 10.1063/1.3585980

[110] Chirita M, Sooryakumar R, Xia H, Monteiro OR, Brown IG. Observation of guided longitudinal acoustic modes in hard supported layers. Physical Review B. 1999;60:R5153-R5156. DOI: 10.1103/PhysRevB.60.R5153

[111] Ferrari AC, Robertson J, Beghi MG, Bottani CE, Ferulano R, Pastorelli R. Elastic constants of tetrahedral amorphous carbon films by surface Brillouin scattering. Applied Physics Letters. 1999;75:1893–1895. DOI: 10.1063/1.124863

[112] Beghi MG, Ferrari AC, Teo KBK, Robertson J, Bottani CE, Libassi A, Tanner BK. Bonding and mechanical properties of ultrathin diamond-like carbon films. Applied Physics Letters. 2002;81:3804–3806. DOI: 10.1063/1.1510179

[113] Berezina S, Zinin PV, Schneider D, Fei D, Rebinsky DA. Combining Brillouin spectroscopy and laser-SAW technique for elastic property characterization of thick DLC films. Ultrasonics. 2004;43:87–93. DOI: 10.1016/j.ultras.2004.03.006

[114] Dellasega D, Russo V, Pezzoli A, Conti C, Lecis N, Besozzi E, Beghi M, Bottani CE, Passoni M. Boron films produced by high energy Pulsed Laser

Deposition. Materials and Design. 2017; 134:35–43. DOI: 10.1016/j.matdes.2017.08.025

[115] Besozzi E, Dellasega D, Russo V, Conti C, Passoni M, Beghi MG. Thermomechanical properties of amorphous metallic tungsten-oxygen and tungsten-oxide coatings. Materials and Design. 2019;165:107565. DOI: 10.1016/j.matdes.2018.107565

[116] Rossignol C, Perrin B, Bonello B, Djemia,P, Moch P, Hurdequint H. Elastic properties of ultrathin permalloy/alumina multilayer films using picosecond ultrasonics and Brillouin light scattering. Physical Review B. 2004;70:094102. DOI: 10.1103/PhysRevB.70.094102

[117] Alonso-Redondo E, Belliard L, Rolle K, Graczykowski B, Tremel W, Djafari-Rouhani B, Fytas G. Robustness of elastic properties in polymer nanocomposite films examined over the full volume fraction range. Scientific Reports. 2018;8:16986. DOI:10.1038/s41598-018-35335-1

[118] Abadias G, Colin JJ,Tingaud D, Djemia Ph, Belliard L, Tromas C. Elastic properties of α- and β-tantalum thin films. Thin Solid Films. 2019;688: 137403. DOI: 10.1016/j.tsf.2019.06.053

[119] Bottani CE, Li Bassi A, Beghi MG, Podesta A, Milani P, Zakhidov A, Baughman R, Walters DA, Smalley RE. Dynamic light scattering from acoustic modes in single-walled carbon nanotubes. Physical. Review B. 2003;67: 155407. DOI: 10.1103/PhysRevB.67.155407

[120] Polomska AM, Young CK, Andrews GT, Clouter MJ, Yin A, Xu JM. Inelastic laser light scattering study of an ordered array of carbon nanotubes. Applied Physics Letters. 2007;90: 201918. DOI: 10.1063/1.2741145

Index

A

Accessible, 112, 115, 121, 123

Account, 44, 118-120, 122-123

Acoustic, 106-115, 117-120, 122, 124, 126-133

Adhesion, 3-4, 104, 109

Agreement, 6, 40, 68, 79, 93, 97, 99

Algorithm, 110

Aluminum, 12, 118

Amorphous, 6, 9, 112, 123, 126, 132-133

Analogous, 14, 16, 47, 69, 72, 108, 112, 116

Analogy, 13-14, 16, 27, 55

Analysis, 9, 32, 64, 90

Angular, 43, 55, 67-68, 77-80, 82-84, 105, 108

Antennas, 120

Architectures, 103

Asymptotic, 13, 15, 28, 33-34, 38, 41, 43-47, 49-50, 52-53, 55, 57, 60, 65, 67, 77, 79, 88

Atomic, 1-2, 5-6, 105-107, 113, 117, 128

Atoms, 1, 3, 5-9, 99, 105-106

B

Backscattering, 123

Bandwidth, 112, 116

Biomedical, 125, 132

Borosilicate, 118

C

Carbon, 99, 107, 112-113, 117, 123, 125-127, 132-133

Cardiac, 103

Ceramics, 121, 123

Coatings, 2-3, 5, 122, 133

Coefficients, 29, 40-44, 47, 49-50, 53-54

Compatibility, 18, 21, 24, 28, 37, 40, 42, 53

Component, 17, 27-28, 45, 72, 77, 108-110, 113, 115, 119, 122

Configuration, 15, 105, 119-121, 123

Crystal, 91, 96, 99, 119-120, 130

Crystallographic, 93-94, 99, 102

Cylindrical, 16, 19, 28, 38, 76, 114, 116, 121, 125

D

Deformation, 14, 17, 68-69, 88, 105-106, 121

E

Equation, 13-14, 18-21, 24-25, 47-48, 53, 62, 71, 89, 94

Equilibrium, 14, 18-24, 28, 36, 40, 42, 53, 68, 105-106

Experimental, 1-2, 4, 6, 106, 113, 115-116, 118, 121, 129

F

Fabricated, 120, 127

Factor, 4, 34, 56, 72-75, 78, 80-86, 118
Ferroelectric, 128
Framework, 14, 17, 67
Function, 14-16, 26-27, 33-34, 39, 68-70, 72, 74-75, 94, 106, 113

G

Gallium, 7-8
Gasochromic, 104
Geometry, 73-74, 84-85, 93, 111, 114-116, 120, 122-123
Glycol, 104
Graphite, 92, 96-98, 102

H

Harmonic, 107, 122
Heterodyne, 121, 131
Hexagonal, 93, 96, 104, 110, 120
Homogeneous, 62, 95-96, 99, 106-107, 109, 118-119, 125
Hybridization, 92
Hydrothermal, 92, 94, 99

I

Inhomogeneous, 125
Interatomic, 106-107
Isotropic, 13-16, 26, 107-108, 130

J

Journal, 11-12, 32, 64-65, 90, 127, 129, 131

K

Kinematical, 17, 38, 52

L

Laser, 2, 112, 116, 121, 127, 130-132
Lattice, 93-94, 120
Leopold, 126
Linear, 13-14, 32, 40, 77-78, 87, 90, 108-111
Lithographic, 113, 119

M

Magnetic, 104
Magnitude, 114, 119
Matrix, 3, 16, 21, 120
Mechanics, 12, 32, 64-65, 90, 125
Membranes, 113, 119, 123, 126
Mesoscopy, 12
Metallic, 2-4, 119, 130, 133
Micrometers, 114, 117, 120, 122-123
Microscopes, 112
Microstructure, 128
Models, 13-14, 105, 123
Monochromatic, 107
Monoclinic, 93-94, 96-97, 99, 102, 104
Monocrystalline, 120
Morphologies, 91-92, 95
Motion, 105-106, 111-113, 115, 117, 122, 126
Multilayer, 5-6, 12, 118, 129, 133

N

Nanocomposite, 1, 4, 96-103, 133
Nanocrystalline, 113, 117, 122-123, 126-128, 131
Nanoparticles, 1-4, 7-8, 92, 104, 120, 130
Nanoplate, 128
Nanopowders, 104
Nanoribbons, 103
Nanosheets, 95, 97, 99, 102, 104
Nanotechnology, 11-12, 124, 126, 129-130

Nanotubes1, 107, 113, 123, 126, 133
Nanowire, 120, 126, 130
Nonclassical, 24, 36-37, 53
Nonlinear, 110-111

O

Optomechanics, 119, 122, 129
Oscillations, 105, 107, 111, 118, 128
Oxidation, 92-93, 102
Oxygen, 96-97, 99, 133

P

Parallel, 96-97, 108, 110, 114, 119-120, 122
Penetrates, 111, 122
Periodic, 106, 108-110, 115, 118-119, 121-122, 131
Phonons, 106, 119, 125, 128-129, 131
Photocatalysts, 91
Photonic, 119-120, 129-130
Polarization, 107, 109, 130
Polymer, 103, 116, 133
Polymorphs, 96, 104
Pristine, 116, 120
Propagates, 114, 116
Prototype, 109
Pulse, 3, 114-121, 128

Q

Quantum, V, 1-2, 6-7, 112, 125

R

Rectangular, 76, 110, 125
Remarks, 28, 60, 68, 88
Resonators, 105, 111-113, 126

S

Segments, 123
Semiconductor, 91, 130
Sensor, 103, 112, 126
Silicon, 7-8, 127-128
Sodiumoxide, 103
Superlattice, 118-119, 125
Symmetric, 14, 16-17, 29-30, 106, 110, 116
Symmetry, 13, 15-17, 26-29, 31, 48, 50, 68, 79, 82, 88, 108, 123

T

Temperature, 3, 7, 92-94, 99, 102, 112-113, 117-118, 121
Tension, 67-68, 73, 88, 112-113
Tetrahedral, 112, 123, 126, 132
Thermoelasticity, 64
Thermoplastic, 128
Trioxide, 91

U

Ultrafast, 125, 127-128

V

Vacuum, 11, 127
Validate, 67, 72, 92, 120
Validation, 75
Velocity, 108-109, 111, 113, 117-118, 121, 128
Viscoelastic, 13-14

W

Wavelength, 106, 108-109, 111, 114, 117-120, 123